AUTO SAFETY:

ASSESSING AMERICA'S PERFORMANCE

AUTO SAFETY

Assessing America's Performance

JOHN D. GRAHAM
Harvard School of Public Health

AH *Auburn House Publishing Company*
Dover, Massachusetts

Library of Congress Cataloging in Publication Data

Graham, John D. (John David), 1956–
    Auto safety : assessing America's performance / John D. Graham.
      p.   cm.
    Includes index.
    ISBN 0-86569-188-6
    1. Automobiles—Safety measures—Government policy—United States.
I. Title.
HD9710.U52G67   1989
363.1′256—dc19   88-26700
                                             CIP

Printed in the United States of America

CONTENTS

CHAPTER 4
The Nixon White House and Ford's "Better Idea"   60

CHAPTER 5
Negotiating Safety   88

CHAPTER 8
The Master Pluralist   169

CHAPTER 9
A New Strategy at Ford   195

PREFACE

America has struggled for thirty years to resolve controversy about the use of occupant restraint systems in motor vehicles. Using democratic pluralism as a descriptive framework, the book explains why it was so difficult to resolve the dispute and how the breakthrough toward resolution was ultimately achieved. More general insights are provided for policymakers, corporate executives, lobbyists, consumer activists, attorneys and judges, and legislators who are participants in regulatory processes.

Since there is no such thing as a completely neutral historical account, the reader has a right to know how the author became interested in this topic, what prejudices the author may have, how the work was conducted, and what the sources of financial support were. In the mid-1970s, as a national college debater, I advocated enactment of both mandatory air bags and mandatory seat belt wearing laws. Later, at Carnegie-Mellon University, I wrote a dissertation on this topic from an economic point of view. Despite my prior advocacy and analysis, I have attempted to write the history of the controversy in a fashion that gives credence to a wide range of policy viewpoints and institutional perspectives. Since I was not a participant in the history, I have relied upon the interviews of participants and written source materials described in the appendix. Financial support for this work was provided by the New England Injury Prevention Research Center in the form of a grant to the Center from the U.S. Centers for Disease Control.

The manuscript itself was the product of an embarrassing number of iterations. I received especially helpful comments from Terry Davies, Sheila Jasanoff, Ilana Lescohier, John Romani, Roger Solt, and Tod Woodbury. Many of my interview sources (see the appendix) also provided extensive comments on earlier drafts of the manuscript. Obviously, none of these critics bear any responsibility for what is in the book, and some of them have made clear that they do not agree with all of my conclusions. As a gesture of respect to all who participated directly in the complicated thirty-year history, I want to emphasize that the history I have written is a reconstruction based entirely on the views of others. Hopefully my distance from the fray is more a source of objectivity than of ignorance.

I conclude by acknowledging the tremendous degree of institutional support I have received at the Harvard School of Public Health. I am especially grateful to Bob Blendon, Harvey Fineberg, Marilyn Graham, Bernie Guyer, Fred Mosteller, Marc Roberts, and Milton Weinstein. Without the encouragement and support of these people, it is hard to imagine that I would have pushed this effort to a conclusion.

Boston, Massachusetts   JOHN D. GRAHAM
November 1988

AUTO SAFETY:

ASSESSING AMERICA'S PERFORMANCE

Chapter 1

THE PLURALIST PERSPECTIVE

Modern American pluralism is the amalgam of capitalism, expansive governmental authority, and interest-group politics that has characterized the United States since the 1930s.[1] The central tenet of democratic pluralism is that there should be no single source of sovereign power but rather numerous sources of power which cannot be completely consolidated.[2] Distribution of power is obviously important because people have conflicting wants and values. Our founding fathers were attracted to constitutional pluralism for different reasons: James Madison, for instance, saw pluralism as a strategy to protect aristocracies from a tyranny of the majority, while John Adams saw pluralism as a defense of ordinary people against tyranny by aristocracy.

In modern times the more ambitious claim has been made that democratic pluralism allows for an activist and beneficial government.[3] Unrestrained private capitalism has been replaced by a mixed public-private system in which capitalistic actions are constrained in various ways by regulatory authority. The rise of regulatory authority over private business is still pluralistic because the authority is somewhat fragmented and subject to numerous checks and balances. Regulatory power is also influenced through various channels by the pressures and informal bargaining of interest groups.

This book examines how modern American pluralism operates in the context of a rich case study: the struggle for safer automobiles. The book is primarily a descriptive history. No attempt is made to test pluralism as a scientific hypothesis or to evaluate pluralism as an approach to societal decision making. Although readers may nonetheless form some opinions on these topics based on the his-

1

tory, the pluralist perspective is used solely as a framework to organize the complex thirty-year saga.

Taking American pluralism as a given, the book also examines in some detail the performance of key actors in the drama. The focus is on two particular power centers: the leadership of the federal government's regulatory apparatus and the leadership of the domestic vehicle manufacturers. Although other power centers also play significant roles in the history, these two power centers were arguably the most central to the saga. In any event, a central hypothesis of the book is that the most "successful" (i.e., influential) actors in the history are those whose strategies and tactics took account of the pluralist nature of American society.

As a background for the detailed chronology in Chapters 2 through 9, this chapter provides the following:

- Some background on the evolution and nature of America's auto safety dilemma.
- A brief definition of the two key elements of pluralism, namely, conflicting wants and multiple power centers.
- Some general approaches that federal regulators and corporate executives might employ when confronted by the auto safety issue in a pluralist society.

The empirical content of the book centers on America's struggle since the 1960s to employ lifesaving technologies such as automobile lap and shoulder belts and inflatable air cushions ("air bags").

America's Safety Dilemma

The post–World War II era brought Americans unprecedented prosperity and mobility. Automobiles are now a central feature of American culture, reflecting several decades of rising per capita incomes, the suburbanization of cities, and the construction of an immense interstate highway system. Americans have come to love their cars, which serve as a symbol of status and accomplishment as well as a private transport vehicle.

The motorization of America is reflected in the most basic transportation statistics. The number of motor vehicles registered in the United States increased from 31 million in 1945, to 89 million in 1965, to 163 million in 1985. The five-fold increase in vehicle registration from 1945 to 1985 far outpaced the 82 percent increase in the overall size of the U.S. population during this period.[4] The data on actual miles of motor vehicle travel tell an even more dramatic story. Americans traveled 250 billion vehicle miles in 1945, 888 billion vehicle miles in 1965, and 1.7 trillion vehicle miles in 1985.[5]

America has paid a price for enhanced mobility. Substantial fractions of the country's labor and material resources have been devoted to transportation-related activity. A less obvious component of the price has been the large number of people injured, crippled, or killed in motor vehicle crashes. The National Safety Council reports that national counts of traffic fatalities increased from 28,100 in 1945 to 49,000 in 1965. After rising to peak counts near 55,000 in 1972 and 1979, the death toll declined to 46,000 in 1985.[6] The size of the annual mortality problem is staggering, especially considering the fact that the average age of death for injured motorists is 35 years. Indeed, motor vehicle fatalities are the number one cause of death among Americans between the ages of one and 35.

For each American killed in car crashes, several dozen more suffer debilitating nonfatal injuries. Hospital emergency rooms treat crash victims for lacerations and severe bruises on a daily basis. Crash injuries are also a major cause of serious head injury and are the leading causes of paraplegia and quadriplegia. Estimates of the direct and indirect economic costs of crash injuries to society approach $50 billion per year.[7]

America faced a difficult challenge: How could the traffic injury problem be controlled without compromising the benefits of a motorized society? One answer to this question which America has pursued is the design of motor vehicles that are more forgiving to occupants in crashes.

Beginning in the late 1950s, America created two technical choices in its quest for enhanced safety in the "second collision"— the crash of an occupant into the vehicle interior. The most fundamental technical issue pitted "manual" or "active" occupant-protection systems against "passive" or "automatic" systems. The former approach relied on conscious protective behavior by motorists; the latter sought technological innovation to provide crash protection without requiring action by people.[8]

Advocates of the active approach were impressed by the injury-mitigation potential of manual lap and shoulder belt systems. Once such equipment was installed in all new cars, the key problem became the public's widespread nonuse of available systems. The result was advocacy of various use-promoting policies ranging from information and education, to economic incentives, to coercive belt designs, to compulsory belt use legislation.

Advocates of automatic crash protection were optimistic about the creative capacities of the engineering profession. They were intrigued by various passive protection concepts, such as self-fastening seat belts, interior padding and blankets, and inflating air cushions. The principal advantage of these approaches is that they would circumvent the behavioral obstacles to crash protection that had limited the effectiveness of manual restraint systems. Their principal disadvantage was that cars would have to be redesigned to accommodate these untested and possibly dangerous devices, a process that would prove to be both costly and time-consuming.

From a technical point of view, the active and passive restraint systems arose from the same biomechanical principles of crash protection. An occupant must be kept inside the vehicle during a crash if severe injury is to be prevented. Moreover, the occupant's motion during a crash must not bring him or her into violent contact with sharp or hard surfaces within the vehicle interior. Finally, the crash forces that do strike the occupant need to be distributed on the body in such a way that minimizes the probability and severity of injury. These basic principles of crash protection had to be satisfied by auto engineers at the same time that cars were designed to meet customer demands for comfort, convenience, appearance, and performance and new regulatory demands for emission control and fuel economy. Last but not least, the price of the car must be affordable to the American consumer.

Conflicting Values, Ideologies, and Interests

America's pursuit of vehicle crashworthiness proved to be controversial in part because of differing technical judgments about which automobile safety technologies were feasible and appropriate for widespread installation in motor vehicles. This book explores these disputes. The pursuit of safety through crashworthiness was made much more controversial by conflicting values, interests, and ideologies among interested parties. Although it is traditional in American culture to describe disputes in terms of means rather than ends, the auto safety issue exposed some sharply conflicting ends that motivated the behavior of interest groups.

Several cherished values—each with a respectable history in American culture—were portrayed as paramount in policy debates. The public health ethic called for the adoption of whatever technologies and policies would do the most to lessen the incidence of injuries and fatalities on America's roads and highways. These advocates viewed safety as the paramount value. Advocates of free choice, viewing freedom as the fundamental value, argued that policies should be fashioned to respect and enhance the personal freedom and privacy of American motorists. Professional economists pushed certain notions of rationality in policy choice, such as the policy or technical solution dictated by maximizing the difference between monetary benefits and costs. Finally, various (and sometimes conflicting) notions of equity were argued to be critical, such as those Americans who benefit should pay, and each motorist should be guaranteed a minimum level of crash protection.

The pursuit of particular values often leads to emotional attachment to certain ideological principles that in turn have a life and influence of their own. As a result, this book describes passionate pleas for policies on the grounds that they are pro- or antiregulation, pro- or antibusiness, pro- or anticonsumer, pro- or antifreedom, and so forth. As we shall see, adherence to these ideological principles does not always advance the supposedly derivative values. Advocates nonetheless sought to advance their ideological interests by influencing the political outcomes.

Overlying and underlying the values and ideologies are powerful commercial interests. Some automakers, for example, may seek to enlarge their profits by expanding the overall size of the car market. More importantly, each manufacturer seeks a competitive edge against rival manufacturers, both domestic and foreign. Competition in the auto industry has historically been hampered by the concentration of market power in the hands of General Motors and Ford, the two largest firms, and their fear of antitrust action. The Volkswagen Beetle and, more recently, the Japanese imports "solved" this problem with a brutal dose of competition. In this competitive struggle, new safety devices can be viewed as either a marketing edge or a disadvantageous cost factor. Insofar as safety features influence (positively or negatively) new car sales, the vast dealer network also has a material interest in the technological and policy outcomes.

Market forces also influence the motives and behavior of suppliers to the auto industry and insurers of motor vehicles. As air bag suppliers, for instance, seek to expand their market, seat belt makers may feel threatened and take steps to protect and enlarge their business. Meanwhile, casualty insurers may see monetary gain in the mitigation or reduction of crash injuries. Use of restraint systems may reduce the insurer's obligation to make payments to policyholders. Because insurance premiums might never decline or may decline only several years after payouts fall, a windfall for insurers might be anticipated. And, as always, there are America's plaintiff lawyers seeking plausible clients and deep pockets to sue. Not surprisingly, they find the auto injury issue an attractive prospect.

Commercial interests become national political interests once it is clear that public policy decisions will influence the expenditure of billions of dollars. Yet political interests are not always tied to commercial gain. For example, personal political interests emerge once advocates, politicans, and bureaucrats take public stands on controversial issues. These actors develop strong interests in particular resolutions of the auto safety issue when it is apparent that the final outcome will affect their reputation in the eyes of others and in history books. Likewise, organizations such as legislative committees and consumer advocacy groups develop political interests that do not necessarily arise explicitly or directly from potential material gain. An organizational "win" is not always accompanied by monetary reward to leaders or members. Interestingly enough, the most obvious form of political interest—Republican or Democratic advantage—appears to have had only a modest influence on the evolution of the auto safety issue.

Finally, there is the average American's love of the private automobile. When Detroit or Washington does something that affects people's cars or the way people perceive or use them, a vigorous public reaction can be anticipated. Because people who vote also buy cars, there isn't a politician in America who doesn't recognize a sort of populist interest in the right of people to do as they please with their cars.

Power Centers in America

Who in America has the power to determine what decisions will be made about auto safety? The answer to this question is remarkably complex, reflecting the multiple power centers that characterize America's pluralist model of decision making. A sketch of an answer is provided here, to be followed by a richer contextual answer in Chapters 2 through 10.

Automobile manufacturers and consumers were once the sole sources of power in the design of vehicles. The corporations, hierarchical as they are, still exhibit some diffusion of power. While each vehicle manufacturer has its own managerial peculiarities, decisions about a major safety feature are normally analyzed by some combination of the corporation's production, engineering, legal, finance, marketing, research, government affairs, and public relations staffs. Final decisions about major new safety features are normally made by the firm's top management, the president and/or chairman of the board.

Before enactment of federal legislation in 1966, state governments exercised their power to regulate the safety of vehicles sold in their states. Many of the original seat belt equipment laws in the

United States were actually enacted by state legislatures. When the federal government assumed primary authority over vehicle safety in 1966, it simultaneously preempted the authority of state legislatures to regulate vehicle safety independently. Despite this usurpation of state authority, states retained primary responsibility for regulating the use of safety features by motorists and determining the content of most traffic laws. Even so, the federal government possesses power to influence state traffic laws and highway designs by threatening a cutoff in federal highway construction funds or by offering incentive grants.

At the federal level, power over vehicle design is distributed among the legislative, executive, and judicial branches. There are few limits on the power of Congress to regulate the auto industry on safety matters. In practice, Congress has delegated authority to an administrative agency within the executive branch. The Department of Transportation (DOT) can now devise safety standards and compel manufacturers to meet them. Congress retains control over the administrative agency through the appointments, funding, and oversight processes.

Beginning in the 1960s, Congress began to assert more control over some agencies by creating a power of "legislative veto," and enacted such power over DOT in a limited area in the 1974 amendments to the 1966 Safety Acts. If Congress objected to a DOT safety rule, a simple majority vote of both houses (without a presidential signature) could overturn it. In 1983 the U.S. Supreme Court ruled unconstitutional this power of legislative veto because it bypassed the President.[9] Today, an authorized administrative rule can be overturned by Congress only if new legislation is passed and signed into law by the President (or if a presidential veto is overridden).

The administrative agency is not the only source of power within the executive branch that can determine the safety of vehicles. The leaders of administrative agencies in the executive branch are appointed to office by the President (with Senate approval) and can be fired at will by the President. The White House also exerts control over executive agencies through budgetary requests and executive orders. In recent years the Office of Management and Budget (OMB) has played an instrumental role in reviewing (i.e., blocking, delaying, and modifying) agency actions, a role authorized by presidential executive order.

An administrative agency's rule that is reviewed favorably by the White House and is not overturned by the Congress does not necessarily become law. The rule may be challenged by its opponents in court, where federal judges possess power to review agency rules. Federal courts may overturn such rules if they are determined to be "arbitrary, capricious, an abuse of discretion or otherwise not in accordance with the law."[10] The key factors in administrative litigation are supposed to be the agency's "statutory mandate" (letter of the statute and legislative history) and the existence of a "record" containing substantial evidence in support of an executive agency's decision. Courts can also block agency action if proper procedures have not been followed (e.g., failure to hold public hearings or to provide timely public notice of a new rule). Courts also review agency decisions not to issue rules or to rescind rules. The parameters for judicial review of agency action and inaction were enumerated by Congress in the Administrative Procedure Act of 1948 and have been embellished through case law.[11]

In summary, because multiple power centers characterize modern American pluralism, there are many veto and originating points for an auto safety rule. This fragmentation of authority, rooted in the Constitution, has been amplified by subsequent legislation and judicial decisions. The multiple sources of power are intended to provide each interest group with ample opportunity to influence the formulation of safety policy.[12]

Can Pluralism Work?

Democratic pluralism has a long and respectable intellectual history, beginning with the explicit intentions of the framers of the Constitution. The system was designed to make bargaining, negotiation, and compromise the essential features of the American political style.[13] Without give and take among conflicting interests nothing could be accomplished, as each interest group could exercise its veto power through one of the numerous power centers.

Advocates of pluralism make some remarkable claims. They say that pluralistic competition among interest groups will tend toward an equilibrium, that the operation of pluralism is automatic, and that the public policies arising from pluralistic equilibrium will approximate the overall public interest.[14] In particular, they insist that no central planning or authority is necessary to achieve the public interest.[15]

The claims on behalf of pluralism are unfortunately not falsifiable (testable), because concepts such as "the public interest" and "pluralist equilibrium" are not self-defining. Rather than examine an untestable hypothesis, I simply will describe how pluralism produced a resolution of this issue in the United States. I leave it to others to judge how well democratic pluralism worked and whether some other political-economic system might have done "better." Chapter 10 advances some possible incremental reforms of America's pluralist approach to safety regulation that might have led to a swifter, more competent, and more democratic resolution of the controversy.

Managerial and Administrative Strategy

This book goes beyond a pluralist interpretation of the struggle for auto safety and looks at the same issue from the perspectives of both corporate executives and administrators of federal agencies. Because leaders of private corporations and regulatory agencies can do little to alter the basic structure of American society, they must pursue their personal and organizational objectives within the confines of America's political-economic system. Managerial and administrative strategies must therefore be devised and implemented on the basis of a keen understanding of the realities of modern American pluralism.

To explore such strategic issues, this book examines in some detail the personalities of key leaders in the auto safety saga, the strategic dilemmas and tactical options they faced, and the choices they made. Likewise, an effort is made to describe the ambitions and strategies of those who proved to be the adversaries of corporate and regulatory leaders. This book does not attempt to question the values and goals of these actors but instead examines whether the strategies they adopted were consistent with their professed values and goals. To draw some general insights from such a rich historical drama, it may be useful to consider some general approaches that business executives and regulators might take to respond to the auto safety issue.

Strategic Corporate Management

The goal of strategic management is "to position and relate the firm to its environment in a way which will assure its continued success and make it secure from surprises."[16] "Success" of for-profit enterprises in a capitalist system is readily definable. As a first approximation, success means serving stockholder interests through profit maximization, although there are some tensions between, say, short-run and long-run profit.

Since 1950 the challenges facing the leadership of corporate America have been exacerbated by factors such as "increasing global competition and large scale societal involvement in determining how the firm is to be run and the role it should play."[17] Rapid technological progress further complicates the picture, as corporate leaders must decide when and how fast to discard one technology in favor of another. At the same time, the corporate executive faces the traditional yet fundamental problems of production, marketing, and financial control in an environment where consumer demands may be changing rapidly yet unpredictably. As we shall see, there is no better illustration of the corporate strategy dilemma than the environment faced by the domestic automobile industry during the 1970s and 1980s.

Strategic planning has been proposed—and is often employed—to help resolve the corporate executive's major challenges. "Planning" is defined as "the core capacity developed by firms to adapt to environmental movements."[18] Rather than passivity and reaction to external forces, strategic planners see "adaptability" as "an active, creative, and most decisive search for the conditions that can secure a profitable niche for the firm's business."[19] By anticipating changes in the firm's external environment, corporate executives can rapidly deploy internal resources to improve its competitive position.

What then are the implications for corporate strategy of a potentially powerful Washington-based regulatory apparatus that is controlled by politicians in a pluralist process? At the risk of oversimplifying answers to a complex question, one can imagine at least three generic responses to the rise of regulatory authority.

The first response is the *obstructionist view*—which continues to have a profound influence in corporate America—and advocates that firms should band together in an attempt to undermine the political legitimacy of regulatory action. To accomplish this end, firms support both broad-based educational efforts to promote free-market ideology and specific political and legal tactics to render impotent the powers of regulatory agencies. The obstructionist view sees firms within one or more regulated industries as allies in a war between the public and private sector.

The second response, what might be called the *passive acceptance view*, sees the regulatory agency as a legitimate yet exogenous environmental influence that each firm in an industry learns to accept. Once regulatory policies are put in place, firms expend only the minimum amount of resources necessary to comply with specified rules. Compliance costs are viewed as a uniform price that every firm in the industry must pay to be in business. In short, the response is to accept regulation, cooperate, and comply.

Finally, the *opportunistic view*, in contrast, sees the regulatory agency as either a potential resource or a liability in the firm's competitive struggle for survival and prosperity. Bluntly stated, the firm conceives the powers of the regulator as a highly effective weapon during competition with other firms in the same industry. The challenge is to effectively manipulate regulatory powers to serve the firm's interests while preventing competitors from doing the same.

Of course, these three views of corporate strategy in a regulated environment are not mutually exclusive. Indeed, it is not uncommon to observe the same management team employ all three courses of action simultaneously in various aspects of corporate policy. In the auto safety saga, we will observe all three of these responses employed at particular times.

The implications of regulatory power for corporate policy are so profound that firms must reorient their thinking and invest substantial resources in new adaptation capability. As a starting point, strategic planners argue:[20]

> *Managers need to understand ideologies other than free enterprise, recognize their legitimacy, and be prepared to change from a traditional* dogmatic *belief that free enterprise is the only "truth" to a more open* ecumenical *position which accepts the legitimacy of other ideological "truths."*

Even adherents to the obstructionist view must appear open-minded toward alternative ideologies, because the public forums where they will attempt to educate people will place a premium on perceived openness to new ideas.

In addition to learning other points of view, strategic planners urge corporations to acquire an enlarged information communication system covering the social and political as well as commercial spheres of life. To do so, corporate managers must learn the language and cultures of other spheres so that they can translate information conveyed in these languages into the traditional parlance of business communication.[21]

Finally, strategic planners emphasize the importance of understanding political processes and developing skills appropriate for political behavior. The contention is that adapting to regulatory challenges is very different from dealing with business opportunities. The business world is based primarily on consensual, uni-ideological decision processes, whereas the regulatory world is comprised of conflict-oriented, multi-ideological decision processes in which no party has decisive power.[22] Firms need to identify their political marketplaces and engage in lobbying, coalition formation, bargaining, and the use of propaganda. Firms may also need to hire people with very different skills and devise ways to evaluate and promote them. Whether this capacity should be developed in-house or acquired through contractual arrangements with outside parties is a key question that has no universal answer.

In summary, the rise of regulatory power over business in conjunction with fierce global competition and rapid technological change creates an enormous challenge for the corporate executive. Strategic management is a potential procedural solution to this challenge that emphasizes planning, adaptability, and sophistication in dealing with government. In Chapters 2 through 9, we will examine how numerous firms, large and small, responded to the rise of Ralph Nader and his regulatory legacy.

Styles of Administrative Management

The leader of a federal regulatory agency is typically responsible for administering laws that contain ambitious congressional goals, authorize the use of particular policy tools, provide very limited resources, and demand immediate action. At the same time, this leader is typically a political appointee with strong ties to certain actors or philosophies within a presidential administration and therefore has a projected tenure unlikely to exceed several years. The activities of the "administrator" are likely to be under close scrutiny by both congressional patrons and interest groups that might be affected by policies of the agency.

In this book, we examine the roles of two political offices, the secretary of transportation and the administrator of the National Highway Traffic Safety Administration (NHTSA), as they relate to achieving congressionally specified safety goals. We focus primarily on the use of various styles of administrative management to accomplish congressional objectives while advancing the leader's personal and political interests.

"Style of administrative management" refers to the ways in which a leader chooses to utilize an agency's resources and authority to accomplish both personal and congressional goals. In particular, the approach taken to influence the technological choices of regulatees—in this case vehicle manufacturers—is a fundamental aspect of administrative style. It may be useful to consider three conceptually distinct administrative styles that a DOT secretary or NHTSA administrator might employ.

A *technology-forcing* style, which was clearly envisioned by the architects of auto safety legislation, implies a top-down arrangement in which regulators dictate to firms far-reaching standards that compel development and deployment of new safety technologies.[23] Such a heavy-handed style is believed to be necessary because private firms are considered reluctant and often resistant to take technological risks that might enhance safety. The administrative agency is supposedly blessed with sufficient will and power to withstand such resistance. According to this model, the key impetus for implementing safety innovations comes from the regulatory agency.

A *technology-inducing* regulatory style entails the more subtle use of regulatory powers to encourage one or more firms to see that technological innovation is in their competitive interests. Once one major firm commits to technical change pursuant to—or in anticipation of—regulation, competitive pressure may build for other firms to do the same. Hence, administrators who adopt a technology-inducing style don't view themselves as the primary source of industry-wide innovation. They instead see themselves as a catalyst of competitive forces aimed at enhanced safety.

An increasingly popular administrative style is one that seeks to *negotiate safety*. Regulatory powers are used as a lever or threat to induce interested parties to come to the negotiating table and bargain in good faith.[24] Administrators, serving as either mediators or principals in the negotiations, use regulatory powers only as a last resort when consensual safety agreements cannot be worked out among interested parties. The hope is that the threat of regulatory action will cause corporate executives to conclude that they can do better (or at least no worse) through negotiation than through formal rule making.

Finally, a *nonregulatory style* uses nontraditional administrative tools to advance auto safety without regulation.[25] Such tools might include persuasion, jawboning, demonstrations, subsidized research and development, and tax credits and economic incentives. In some situations the choice of the nonregulatory style reflects a belief that competitive market forces by themselves are sufficient to assure auto safety advances. Alternatively, the consumer demand for safety may be facilitated or enhanced by the provision of information or incentives by the agency.

Some administrative styles are more attractive than others depending upon the leader's ideological orientation, the political spin of a particular safety issue, the financial condition of regulatees, and current political sentiments. Moreover, some leaders are more capable of executing certain styles due to their background, training, experience, personality, and/or political assets and vulnerabilities.

In the chapters ahead, we shall examine in some detail how well these various styles have worked and whether other styles might have worked better in particular contexts.

Endnotes

1. Theodore J. Lowi, *The End of Liberalism* (New York: W. W. Norton & Company, 1969), p. 29.
2. Robert A. Dahl, *Pluralist Democracy in the United States: Conflict and Consent* (Chicago: Rand-McNally and Company, 1967), p. 24.
3. Lowi, *The End of Liberalism*, p. 46.
4. *Motor Vehicles Facts and Figures '86* (Detroit, Michigan: U.S. Motor Vehicle Manufacturers Association, 1986), p. 19.
5. *Highway Statistics Annual* (Federal Highway Administration, U.S. Dept. of Transportation, Washington, D.C.), various issues.
6. *Accident Facts 1984* (National Safety Council, Chicago, Illinois, 1985), pp. 58–59.
7. *Motor Vehicle Safety 1983* (U.S. Dept. of Transportation, NHTSA, Washington, D.C., May 1985), DOT-HS-806-731, Table A-17, pp. A25–26; also see John D. Graham, "Automobile Safety: An Investigation of Occupant-Crash Protection Policies" (Ph.D. Dissertation, Carnegie-Mellon University, 1983).
8. William Haddon, Jr., and J.L. Goddard, "An Analysis of Highway Safety Strategies," in *Passenger Car Design and Highway Safety* (Conference Proceedings, Association for the Aid of Crippled Children, 1962).
9. *Immigration and Naturalization Service* v. *Chadha*, 103 S. Ct. 2764 (1983).
10. Section 706(2) (A) of the Administrative Procedure Act, 5 U.S.C. 706.
11. Section 706(2) (E) of the Administrative Procedure Act, 5 U.S.C.706. See also John D. Graham and Patricia Gorham, "NHTSA and Passive Restraints: A Case of Arbitrary and Capricious Deregulation," *Administrative Law Review* 35:193–252 (Spring 1983).
12. Kenneth M. Dolbeare and Murray J. Edelman, *American Politics: Policies, Power, and Change* (Lexington, Mass.: D.C. Heath, 1971), p. 262.
13. Ibid., p. 263.
14. Lowi, *The End of Liberalism*, pp. 46–47.
15. Charles E. Lindblom, *Politics and Markets: The World's Political-Economic Systems* (New York: Basic Books, 1977), pp. 258–259.
16. H. Igor Ansoff, *Implanting Strategic Management* (Englewood Cliffs, N.J.: Prentice-Hall, 1984), p. xv.
17. Ibid.
18. Arnold C. Hax and Nicholas S. Majluf, *Strategic Management: An Integrated Perspective* (Englewood Cliffs, N.J.: Prentice-Hall, 1984), p. 1.
19. Ibid.
20. Ansoff, p. 149.
21. Ibid.
22. Ibid.

23. Stephen Breyer, *Regulation and Its Reform* (Cambridge, Mass.: Harvard University Press, 1982), pp. 106–107.

24. Philip J. Harter, "Negotiating Regulations: A Cure for Malaise?" *Georgetown Law Journal,* vol. 71 (1981), pp. 1+.

25. Robert F. Poole, Jr. (ed.), *Instead of Regulation: Alternatives to Federal Regulatory Agencies* (Lexington, Mass.: Lexington Books, 1982).

Chapter 2

THE BIRTH OF FEDERAL REGULATION

The pluralist model of politics is designed to make radical reform the exception rather than the norm. This model expects the clash of interest groups at multiple decision points to produce no more than incremental change because the losers from radical change can exert power to block the ambitions of their enemies. Moreover, incremental reforms can be designed to compensate losers directly, or political bargaining can result in gains for losers on other seemingly unrelated issues. Radical reforms by their nature are difficult to adopt and implement without creating a determined, well-organized opposition.

Although pluralism is profoundly conservative in its resistance to change from the status quo, radical reform in America is not impossible. A case in point is the creation of federal regulatory authority over the safety of motor vehicles. As we shall see, this example is so peculiar in its evolution that it serves to illustrate how difficult it is in America to accomplish major political change. The example is also remarkable because, in the final analysis, federal authority over the automobile industry—radical as it was—occurred swiftly and without serious opposition.

The 1966 Safety Acts

On September 9, 1966, President Lyndon Baines Johnson signed into law the National Traffic and Motor Vehicle Safety Act and the Highway Safety Act. These laws placed the federal government in the leadership role of a comprehensive national program to reduce

16

the number of injuries and deaths on America's highways. Together, the two laws covered all facets of the traffic safety problem: the vehicle, the tire, the driver, and the road. Ambitious safety programs were to be developed and administered by the new National Traffic Safety Agency, which was originally housed in the U.S. Department of Commerce.

The most stunning and far-reaching aspect of the new legislation was the power given to the federal government to regulate the auto industry—that is, the authority to establish minimum safety standards for all new motor vehicles sold in the United States. This sweeping assertion of governmental power over the nation's largest industry was passed swiftly and unanimously by the Congress, despite initial opposition from the "Big Four" automakers (General Motors Corporation, Ford Motor Company, Chrysler Corporation, and American Motors Corporation).

A federal role in the design of safer cars may not seem earthshattering today, but it was considered a radical notion at the time. Journalist Elizabeth Drew of the *Atlantic Monthly* wrote in October 1966 that this legislation was "a radical departure from the government's traditional, respectful, hands-off approach to the automobile industry, an industry who politician and businessman alike had long considered sacrosanct."[1] Indeed, the auto industry was notable as the only major sector of the transportation economy relatively free of government regulation. Many skeptics were probably asking—with considerable justification—how politicians and bureaucrats in Washington could help engineers in Detroit design safer cars.

The history of the 1966 traffic safety laws (hereafter called the 1966 Safety Acts) provides more than just a legal context for subsequent policy debates about automobile safety belts and air bags. This history offers an introduction to many of the key ideas, personalities, and interests that influenced auto safety issues for twenty years thereafter. The origins of this legislation are also of general interest because the 1966 Safety Acts are considered a forerunner of several dozen health, safety, and environmental statutes adopted in the 1970s to curb alleged abuses of corporate America.[2]

The Intellectual Groundwork

Prior to passage of the two 1966 Safety Acts, the prevailing public view was that crash-related injuries are primarily a behavioral problem. The driver in particular was seen as the causative agent in most crashes, and much of the scholarly thinking in this area was devoted to assigning fault for legal purposes. GM vice president Harry F. Barr expressed the dominant opinion to a *New York Times* reporter in January 1965: "The driver is most important, we feel. If the drivers do everything they should, there wouldn't be accidents, would there?"[3] The consequence of this thinking was that the "safety establishment" (the National Safety Council, state safety officers, medical organizations, law enforcement agencies, auto companies, and insurers) concentrated on driver-oriented countermeasures, such as driver training, education, and traffic law enforcement.

Pioneers in Crash Injury Research. The intellectual underpinnings of the 1966 Safety Acts can be traced to the pioneers of "crashworthiness" research who began to speak out in the late 1940s. Their idea was that vehicles could be designed to better protect drivers and passengers when crashes occur. Because some crashes are inevitable (even with perfect drivers), the mission of physicians and engineers should be to design a forgiving interior that minimizes human trauma in the crash environment.

One of the earliest pioneers was Dr. Claire Straith, a plastic surgeon based in Detroit. Straith worked with the Detroit police department in 1948 to gather statistical information on injury-producing crashes.[4] She documented the instrument panel as a frequent contact surface and recommended removal of all knobs, cranks, drop-down ash trays, and sharp edges from dashboards and instrument panels. Straith also urged that such hostile items be replaced by rubber crash pads.[5]

At about the same time, a clinician at the University of Virginia hospital, Dr. Fletcher D. Woodward, was beginning to criticize auto engineers for not applying principles of safety that had been established in aircraft design.[6] He noted, for example, that the use of safety belts in cars, "such as are used as standard equipment in aircraft . . . might well be one of the most effective single features in preventing serious injury."[7]

Air Force Colonel John P. Stapp also became interested in car design based on his research about aircraft safety. As a physician and biophysicist, Stapp was interested in the effects of mechanical forces on living tissues and established some of the fundamental criteria for safe aircraft design, aircraft ejection systems, and space flights.[8] At the Aeromedical Facility at Edwards Air Force Base in California, Stapp served as a volunteer in rocket and sled tests to demonstrate that the human body could withstand very high forces if properly restrained and protected. From 1953 to 1956 Stapp conducted research demonstrating how principles of safe aircraft design could be applied to cars.

The efforts of Straith, Woodward, and Stapp were important, but "the most significant developments in crash injury research were largely due to the work of one man, a self-trained engineer and pathologist named Hugh DeHaven."[9] As an aviation enthusiast, DeHaven became interested in making flying safer and later became generally interested in crash injury research. He was intrigued by the frequency of car and air crashes that involved both fatalities and survivors.[10] Throughout the 1940s DeHaven conducted field studies and laboratory experiments that confirmed his theories about the critical role of engineering in mitigating crash injury. By the early 1950s, DeHaven was being contacted by engineers in the auto industry to explain his work. Unlike some of his fellow pioneers, DeHaven was more critical of physicians than auto engineers because he believed engineers could not possibly make major improvements until physicians began to collect scientific data on the causes of injury.[11]

Crash injury researchers did exert a significant influence on industry practice even before auto safety became a national political issue. Chrysler and Ford announced in the spring of 1955 that seat belts would be made available as optional equipment. AMC followed suit, but GM delayed until the next model year. In the mid-1950s GM, Ford, and Chrysler also developed and offered improved door locks to prevent occupant ejection from the vehicle.

On its 1956 models, Ford offered an impressive package of safety options through an aggressive marketing campaign, including seat belts, safety door locks, energy-absorbing steering columns, and padded dashboards. When Ford experienced a major sales loss in 1956, many industry observers concluded that "safety doesn't sell."[12] Whether this inference was correct has been a matter of considerable dispute for decades. In any event, the initial impact of crash injury research on car design was limited, setting the stage for the political issue to emerge in the 1960s.

Political Scholarship. Translating medical and engineering concepts of car safety into a progressive political agenda requires a particular type of scholarship. Such writings must sustain enough technical competence to be credible yet disseminate the central policy message to nonexpert opinion leaders throughout society. In the case of safe car design, the political scholarship was executed brilliantly if not independently by Daniel Patrick Moynihan and Ralph Nader.

Moynihan was an academic and chairman of the New York State Traffic Policy Committee in the late 1950s. In a provocative yet compelling article entitled "Epidemic on the Highway," Moynihan criticized the basic message of the safety establishment.[13] He argued that the prevailing focus on individual responsibility for car accidents "shifts public attention from factors such as automobile design, which we can reasonably hope to control, to factors such as the temperament and behavior of eighty million drivers, which are not susceptible to any form of consistent, overall control."[14] A central theme of Moynihan's prolific writings was that the influence of the auto industry was apparent in the activities of the entire safety establishment, including the President's Highway Safety Committee.[15] Even on the basic issue of data collection, Moynihan exposed such serious weaknesses in the state-of-the-art statistics on highway safety that he concluded it was impossible to assess the overall magnitude of the crash injury problem.[16]

Despite his background as a Harvard intellectual, Moynihan is an inspiring, emotional, and articulate man who manages to exude aristocracy without arrogance. His ambitions for influence on public policy were lofty, but his commitment to traffic safety was not transient. When he joined the Kennedy administration as assistant secretary of labor for policy planning, Moynihan kept a close eye on safety as it arose as an issue of political significance. In 1964 Moynihan hired an energetic Harvard law school graduate named Ralph Nader as a special consultant to help him prepare a report on the federal government's role in traffic safety.

The 30-year-old Nader left his law practice in Connecticut to pursue in Washington a deep interest in car safety begun as a Harvard law student. He worked for Moynihan until the spring of 1965 and then did some consulting for a Senate committee that was orchestrating public hearings on the federal government's role in highway safety. Described at the time as an intense, lanky young man with a quiet sense of humor and an instinct for timing and drama, Nader was also a tireless worker who would type night after night in his rooming house while living off personal savings.[17]

Nader first drew national attention in November 1965 with the publication of *Unsafe at Any Speed: The Designed-In Dangers of the American Automboile*.[18] The book received widespread and largely favorable reviews and was covered as a front-page story by the *New York Times*. Although a centerpiece of the book was a detailed indictment of the safety of a particular car make, GM's Corvair, it had more general themes. It converted statistics on highway deaths and injuries into a documentary of personal tragedies. The auto industry and safety establishment were cast as villains, the latter for its neglect of the "second collision" and crashworthiness research and the former for its preoccupation with horsepower and style at the expense of safety.[19]

Unlike Moynihan, whose contribution to traffic safety was primar-

Table 2–1   Motor Vehicle Deaths, 1950–1968

Year	No. of Deaths	Per 100,000 Population	No. of Vehicles (millions)
1950	34,763	23.0	49.2
1951	36,996	24.1	51.9
1952	37,794	24.3	53.3
1953	37,955	24.0	56.3
1954	35,586	22.1	58.6
1955	38,426	23.4	62.8
1956	39,628	23.7	65.2
1957	38,702	22.7	67.6
1958	36,981	21.3	68.8
1959	37,910	21.5	72.1
1960	38,137	21.2	74.5
1961	38,091	20.8	76.4
1962	40,804	22.0	79.7
1963	43,564	23.1	83.5
1964	47,700	24.9	87.3
1965	49,000	25.3	91.3
1966	53,041	27.1	95.9
1967	52,924	26.7	98.9
1968	55,200	27.6	102.1

SOURCE: National Safety Council, *Accident Facts* (Chicago, Ill., various issues).

ily scholarship, Nader became an energetic political operative on Capitol Hill. By the time Nader's book was published, he was already "engaged in a one-man lobbying operation probably unprecedented in legislative history."[20] In this capacity he worked closely with congressional entrepreneurs and their staffs to place auto safety on the legislative agenda and to fashion a strong, proconsumer bill.

Worsening Safety Problem

Moynihan and Nader were highlighting a social problem, traffic fatalities, that had been gradually worsening since World War II. As the data in Table 2–1 indicate, the number of deaths increased slightly in the 1950s and sharply in the first half of the 1960s. The explanations for these trends are not well understood, but they are believed to reflect an increasingly prosperous economy, the movement of the baby-boom generation into the teenage years, and the increased amount of travel on high-speed roads.[21] Even without the increasing trend, the figure of 40,000 deaths was often cited as a much larger problem than the battle casualties in Vietnam.

A peculiar aspect of the traffic safety problem is that it is clearly procyclical with respect to the economy: Good economic times are generally bad times for highway fatalities.[22] This facet of the issue was distinctive relative to most of the Great Society agenda, which seemed to arise from depressed economic conditions. The political overtones were also bad for the auto industry: When car sales and profits were up, highway deaths were up.

Congressional Entrepreneurs

Passage of the 1966 Safety Act would not have occurred without the intense commitment and hard work of consumer advocates in the Congress. Several members of the Senate and their committee staffs were especially instrumental in getting auto safety on the national agenda and securing the votes necessary to assure passage of reform legislation. Although powerful entrepreneurs ultimately emerged in the Senate, the earliest legislative pioneer was a little-known congressman from Alabama.

Congressman Roberts. On January 5, 1956, Congressman Kenneth A. Roberts (D–Ala.) introduced a resolution in the House of Representatives calling for creation of a new subcommittee to study the problem of traffic safety. A classic initiative by a legislative entrepreneur, the resolution received no active support (or opposition) from either powerful interest groups or the Eisenhower administration. Roberts was on his own.

When the House Rules Committee approved the resolution, Roberts became the first U.S. legislator to chair national hearings on the traffic safety problem. No comparable hearings were held in the Senate in the 1950s, despite some unsuccessful initiatives by Senators Paul Douglas (D–Ill.) and Margaret Chase Smith (R–Maine).[23]

At the first set of hearings in July 1956, Roberts took a conciliatory approach toward the auto industry. By avoiding sensationalism and focusing on technical issues, he hoped to induce voluntary action on the part of the industry to improve vehicle safety. This approach reaped no political results, as industry officials argued that more data were needed before design changes could be justified.

In 1957 Roberts chaired hearings on seat belts but could not marshal much interest in a legislative approach. The next year his committee held hearings on a resolution by Congressman John V. Beamer (D–Ind.) that urged states to take the initiative on auto safety. Although this resolution passed, it achieved no significant safety results. From 1959 to 1961, Roberts concentrated on a bill to require the General Services Administration (GSA) to set safety standards for new motor vehicles purchased by the federal government. This initiative was opposed by both GSA and the auto industry, but ultimately passed the House three times and finally the Senate in 1964.

Just as Roberts was about to pursue passage of more ambitious auto safety legislation, he lost reelection, in November 1964 (after seven terms in the House), to a Goldwater Republican. The remarkable aspect of Roberts's legislative record was his persistent pursuit of auto safety legislation. Indeed, "he stood alone on the issue for nearly a decade."[24]

Senator Ribicoff. Despite the defeat of Roberts, the 1964 elections brought a landslide for the Democrats, who gained two-to-one majorities in both houses of Congress. One of the newly elected Democratic senators was Abraham Ribicoff (D–Conn.), a former governor of Connecticut who had also served as secretary of the Department of Health, Education, and Welfare in the Kennedy administration. Ribicoff saw that Democratic control of the White House and both houses of Congress meant that the time was ripe for passage of major legislative reforms.

Ribicoff was given the chairmanship of the subcommittee on executive reorganization of the Committee on Government Operations. But what issue should Ribicoff tackle? In consultation with aide Jerry Sonosky, he decided to concentrate the subcommittee's efforts on traffic safety. (As governor of Connecticut, Ribicoff was known as "Mr. Safety" because of his highly publicized crackdown on speeding.) After his Senate election, Ribicoff read a review of *Accident Research* (a text by several crash injury researchers) in the *New York Times* and became intrigued by the concept of occupant protection in the "second collision."[25] To justify the subcommittee's involvement on this issue, Sonosky and Ribicoff planned hearings in March 1965 on the *federal government's role* in traffic safety. Sonosky worked with Ralph Nader to help prepare members for the hearings.

The initial round of hearings in March 1965 received little public attention, but Ribicoff went ahead and introduced legislation calling for automobile excise taxes to be reduced only if seventeen safety features required by GSA were installed in all new automobiles.[26] This bill passed the Senate despite opposition from both the auto industry and the Johnson administration; however, the House of Representatives did not pass the bill. The powerful House Committee on Interstate and Foreign Commerce proved to be a bottle-

neck for several pieces of proconsumer legislation in the mid-
1960s. Recall that Congressman Roberts, a senior member of this
committee, had lost his reelection bid in 1964.

Ribicoff chaired a second round of hearings in July 1965, where
he heard testimony from the leaders of the domestic auto industry.
The industry's opposition to Ribicoff's earlier bill had irritated the
senator, who came to the hearing with a series of tough and critical
questions.[27] The drama of corporate America being questioned by a
junior senator attracted journalists and television cameras, which in
turn attracted to the hearing room Senator Robert Kennedy (D–
N.Y.), another member of the subcommittee.[28]

Under questioning from Kennedy, GM President James Roche
acknowledged that the company spent only $1.25 million in 1964
on certain safety functions despite $1.7 billion in profit.[29] The im-
plied neglect of safety by America's largest carmaker received national attention and seemed to turn public opinion in favor of some
form of legislative response.[30]

After the hearings, Ribicoff requested data from each automaker
on the rate of defect-related recalls, and Sonosky compiled them
into an attention-grabbing study of new car defects. When the
study was released that summer, front-page headlines in the *Wall
Street Journal* and other newspapers emphasized an alarming rate
of safety-related defects in new cars.[31] This publicity reinforced the
message of Ribicoff's July 1965 hearings.

Senator Magnuson. Warren Magnuson (D–Wash.) was first
elected to Congress as a representative in 1936 and moved up to
the Senate in 1944. A man with working-class roots and a political
outlook profoundly influenced by the New Deal, he assumed the
chairmanship of the Senate Commerce Committee in 1956, the
committee with primary responsibility for regulation of business.

In 1962, after almost being defeated by a political novice, Magnuson was advised to take steps to strengthen his visibility as a legislative leader.[32] Magnuson's reaction was to shake up his staff and hire
Michael Pertschuk, an energetic attorney, to staff the Commerce
Committee. It was quite natural for Magnuson to gravitate toward
proconsumer legislation because of his deep concern for the welfare of ordinary people.

Magnuson's first move on auto safety was to help Congressman
Roberts pass the GSA bill, which had passed in the House several
times but was going nowhere in the Senate. Roberts sought and
received Magnuson's help in steering the GSA bill through the
Senate.[33] The GSA bill was signed into law by LBJ in 1964 and
resulted in the publication of seventeen preliminary safety standards for the roughly 30 thousand new cars purchased by the fed-

eral government each year. The standards covered crashworthiness features such as anchorages for seat belts, padded dashboards and visors, recessed knobs on the instrument panel, impact-absorbing steering wheels and columns, safety glass, and improved door latches and hinges. Many of these standards amounted to codification of emerging industry practice.

The 1965 Ribicoff hearings on traffic safety achieved national prominence, an outcome that embarrassed Magnuson and Pertschuk. They were annoyed at Ribicoff for usurping the Commerce Committee's jurisdiction.[34] Magnuson made clear that any major auto safety legislation would have to be developed and passed by his committee. The competition between Ribicoff and Magnuson for political credit on this issue was a powerful factor in favor of strong legislation since each senator tried to appear tougher than the other.

Supporting Players. It would be wrong to presume that Roberts, Ribicoff, and Magnuson were the only legislators who were influential in passing the 1966 Safety Act. In the Senate, and to a lesser extent in the House, a small collection of actors played supportive roles.

Senator Gaylord Nelson (D–Wis.) became interested in auto safety when a constituent from Whitefish Bay wrote to him complaining about the lack of safety standards for automobile tires.[35] Nelson was ultimately responsible for tire safety being included in the 1966 Safety Act. He also introduced two bills in 1965 that were later incorporated into the final legislation: One called for the federal government to design, build, and test a prototypical safe car, and the other authorized the secretary of commerce to require that all new cars sold in interstate commerce meet the GSA standards for government-purchased cars.[36]

Senator Vance Hartke (D–Ind.) sought to strengthen federal auto safety legislation by providing for criminal as well as civil penalties for willful violations by industry officials, a provision actively sought by Ralph Nader. Although Hartke lost the criminal penalty proposal, he was successful in adding a number of other strengthening provisions suggested by Nader.[37]

In the House of Representatives the key player was Harley O. Staggers (D–W.Va.), the new chairman of the House Interstate and Foreign Commerce Committee. Prior to Staggers, this committee had developed a reputation among Senate staff for being "defiantly anticonsumer" in legislative matters.[38] Staggers introduced the original Johnson administration bill (described in the next section) in the House, strengthened it in various respects to conform to Senate provisions, and worked out compromises on enforcement and inspection provisions to assure unanimous approval in the House.[39] Another member of Staggers's committee, James A. Mackay (D–Ga.), was influential in insisting that Congress fix responsibility for traffic safety in a single federal agency so that national leadership could be provided.[40]

Position of the Johnson Administration

It is doubtful whether major auto safety legislation would have passed as swiftly as it did without the interest and approval of the White House. As events transpired, LBJ provided more than just passive approval. The Ribicoff hearings caught LBJ's attention and he told special assistant Joseph A. Califano in the fall of 1965 to work up a traffic safety bill as part of a larger legislative initiative on transportation policy.[41]

In LBJ's 1966 State of the Union Address he announced his intention to "propose a Highway Safety Act." The precise content of the bill was disputed within the administration.[42] Califano and Budget Bureau Director Charles Schultze favored a bill *ordering* the Commerce Department to set federal safety standards. The secretary of commerce, John Connor, wanted to *authorize* setting of standards if, after two years, the auto industry had not taken sufficient voluntary action. Connor argued that the threat would induce more safety features than a regulatory imperative that would allow industry to wait for specific regulatory rules.

Connor won the dispute, and the administration introduced in early 1966 the strongest car safety bill yet before the Congress. The initiative was quite a victory for a small group of activists at Commerce led by Dr. William Haddon, Jr. (a physician and ex-colleague of Moynihan's in Albany)—who had been trying for years to persuade the secretary of commerce to take a more activist stance on auto safety. Prior to this legislative initiative, their largest victory had been a departmental position of "not objecting" to standby federal authority to set tire safety standards.[43]

In a speech before the American Trial Lawyers Association on February 2, 1966, President Johnson urged a halt to "the slaughter on our highways," which he described as the "gravest problem before this nation—next to Vietnam."[44] On March 2, 1966, LBJ delivered his message on transportation policy to the Congress. He proposed a six-year $700 million Traffic Safety Act of 1966, which would provide the federal government with a principal policy-making role. He also endorsed a tire safety bill that had been passed by the Senate.

The administration's package was introduced by Senator Magnuson and Congressman Staggers in the Senate and House, respectively. It was soon apparent, however, that this ambitious bill was not ambitious enough. Ribicoff, Magnuson, and Nader were not about to let the administration take the credit for landmark legislation on the safety of automobiles.

As soon as the bill was introduced, Nader labeled it "a no-law law."[45] Senator Ribicoff urged that the bill be strengthened to provide interim standards based on the GSA bill and a requirement that the secretary of commerce set safety standards (instead of the two-year threat).[46] Senator Magnuson opened hearings on the bill by announcing that he would sponsor amendments requiring the GSA standards be applied to all new cars on January 31, 1967, and final standards a year later.[47]

Detroit's Response

During the 1960s engineers in Detroit were not oblivious to the emerging technical interest in crash injury research. The Motor Vehicle Manufacturers Association was supporting biomechanics research at both Wayne State University and the University of Michigan, and Ford and GM had devised new crash test facilities. What Detroit's leadership did not appreciate was the emerging political interest in auto safety.

The auto industry had been largely insulated from Washington politics for decades and was therefore ill-equipped to respond to this accelerating political momentum behind car safety legislation. Elizabeth Drew captured Detriot's provinciality:[48]

> *And back in Detroit, the giant $25 billion-a-year auto industry nestled comfortably in its corporate cocoon, content in the knowledge that this was the pride of private enterprise, the backbone of the American economy (accounting for nearly one sixth of it), and the only major transportation industry so free of government regulation.*

Prior to LBJ's legislative initiative, the industry's position—conveyed through the Detroit-based Automobile Manufacturers Association (AMA)—was that a good job was already being done on auto safety and no new legislation was necessary. After the embarrassing Ribicoff hearings and LBJ's initiative, the industrial forces gave ground grudgingly.

They opposed the administration's bill but offered an alternative. By loosening the antitrust laws, the AMA argued, the manufacturers could get together with the states to set new standards. A modified

Vehicle Equipment Safety Compact was proposed as a device to recommend safety rules to the states. In effect, the industry was seeking to block federal regulation by expanding the role of states.[49] In retrospect, the wisdom of this proposal seems questionable— even from the industry's perspective—since several states (New York, California, and Illinois) were gearing up for passage of ambitious—and possibly conflicting—regulatory programs.

The auto industry's political problems mushroomed in the spring of 1966 when it was revealed that General Motors Corporation had undertaken a private investigation of the background and activities of their outspoken critic, Ralph Nader.[50] GM's legal department had suspected that Nader was financially tied to more than one hundred lawsuits filed against the company in regard to the Corvair.[51] Nader's book, *Unsafe at Any Speed*, had criticized the design of the Corvair and stimulated widespread anxieties among Corvair owners. If Nader could be tied financially to the lawsuits, he could be discredited as an expert witness.[52]

When a preliminary investigation turned up nothing, it was not stopped but expanded into an inquiry into his private life in search of ways to destroy his credibility. Private investigators reportedly followed Nader in Des Moines, Philadelphia, and Washington.[53] According to the *Congressional Quarterly*, Nader was "approached by women and bothered by late-hour telephone calls."[54] Nader also received "reports of agencies inquiring into his background—on the pretense that he was being considered for a new job—with questions on his sex life and whether he was anti-Semitic, a member of a left-wing political group, a licensed driver, and professionally competent."[55]

On February 11, 1966—just after Nader's public testimony before Senator Ribicoff's committee—Capitol police apprehended two detectives who were trailing the lawyer. A month later, on March 9, GM publicly admitted it had conducted a "routine" investigation of Nader.[56] On the advice of attorney Theodore C. Sorenson (a former aide to President John F. Kennedy), GM President James Roche apologized publicly before the Ribicoff committee on March 22.[57] While Roche accepted responsibility for GM's actions, he said the investigation had been initiated, conducted, and completed without his knowledge or the consent of GM's governing committee.

Media Fascination

Even before the publicity surrounding GM's investigation of Nader, the auto safety issue had emerged as an attractive story to journalists and reporters. According to Elizabeth Drew of the *Atlantic*

Monthly, the auto safety issue "had all the makings of a good news story, complete with good guys and bad guys and horror stories, and the press gave it big play."[58] Beginning with the 1965 Ribicoff hearings, where GM executives were skeptically questioned by Senator Robert Kennedy (D–N.Y.) and colleagues, the media became intrigued with the push for auto safety legislation.

Media fascination was inflamed by the competition among political entrepreneurs who each sought credit for championing the auto safety issue. Elizabeth Drew explains how many stories evolved:[59]

> *The widely syndicated columnist Drew Pearson staked himself out as a sentinel against encroachments on a tough bill, and Senators, Representatives, Nader, and Capitol Hill staff members virtually tripped over one another leaking stories to Pearson about who was mangling the bill now.*

The media's search for anticonsumer villains was particularly evident after the massive publicity surrounding GM's investigation of Nader and Roche's public apology before Congress.

The impact of the Nader incident was powerful. It dramatized Nader as a personality, brought attention to his book and ideas, and made it easy for him to command press attention in the future. At the same time, the adverse publicity spilled over from GM to the rest of the industry, undermining what little credibility Detroit had left in the eyes of Washington politicians and opinion leaders.

To cut their losses and get the issue out of the headlines, the AMA in April 1966 hired Washington attorney Lloyd Cutler to direct its lobbying efforts.[60] Cutler had previously worked for the industry on antitrust problems and was reputed to be effective in representing industrial interests in Washington. In 1962, for example, Cutler had represented the pharmaceutical industry before Congress in the legislative battle over drug efficacy and safety legislation. The AMA soon abandoned its opposition to federal regulation and tried to negotiate with Nader and legislators to make the final bill more livable.

Final Legislative Outcome

President Lyndon Johnson signed the National Traffic and Motor Vehicle Safety Act into law on September 9, 1966, just seven months after the administration had submitted its traffic safety proposals to Congress. The final bill was tougher than the original administration bill, reflecting the persistence of congressional entrepreneurs and Ralph Nader. Minor differences between the Senate and House versions were ironed out to the satisfaction of Congressman Staggers and Senator Magnuson under the watchful eyes of Nader, Cutler, and Califano.

On the same day President Johnson also signed into law the Highway Safety Act, a parallel bill aimed at reducing injuries and fatalities through state highway safety plans. In particular, the law required each state to set up a federally approved highway safety program by December 31, 1968, or face loss of 10 percent of federal aid highway construction funds.[61] About $270 million was authorized over fiscal years 1967 to 1969 for grants to establish state programs, with 40 percent of the authorization earmarked for new *local* safety programs. This bill was also tougher than the original administration bill, which had not provided the federal government any significant leverage over states.

Both laws were to be administered by a new federal agency housed in the Department of Commerce. This arrangement was actually temporary because the administration and Congress were simultaneously working on legislation to create a new Cabinet-level Department of Transportation. The safety functions at Commerce were transferred in 1967 to the National Highway Safety Bureau (NHSB) of the Federal Highway Administration, a unit within the new Department of Transportation.

The legal power over the auto industry given to NHSB by Congress was enormous and only broadly constrained. NHSB was authorized by Congress to establish mandatory minimum safety standards governing the performance of new motor vehicles. Each standard must be shown to be "reasonable," "appropriate for the type of vehicle," "practicable," and necessary to "meet the need for motor vehicle safety."[62] Standards were to be written in "performance" terms and were to provide "objective" means of measuring compliance. The legislative history of the act expressed a special congressional interest in "crashworthiness" standards.[63]

Congress expected that standards would be promulgated by NHSB at "the earliest practicable date."[64] In addition to general standard-setting authority, the act required NHSB to establish interim federal motor vehicle standards—based on the seventeen GSA standards—by January 31, 1967, and final standards by January 31, 1968. The overriding purpose of the act was to "reduce traffic accidents and deaths and injuries resulting from accidents."[65]

NHSB's First Director

President Johnson nominated William Haddon, Jr., then special assistant for traffic safety planning to the undersecretary of com-

	Factors		
Phases	Human	Vehicles and Equipment	Physical and Socio-economic Environment
Pre-crash			
Crash			
Post-crash			
Losses	Damage to People	Damage to Vehicles and Equipment	Damage to Physical and Socio-economic Environment

Figure 2-1 The Haddon Matrix.

merce for transportation, to serve as the first director of NHSB. A physician with graduate training in public health, Haddon was an accomplished traffic safety expert who had coedited the influential 1964 textbook entitled *Accident Research*.[66] At the commerce department, Haddon had been working aggressively to extend the federal government's role in traffic safety.

As a researcher, Haddon developed a conceptual framework to aid in thinking about traffic safety countermeasures.[67] The so-called "Haddon Matrix," depicted in Figure 2–1, shows that the time sequence of crashes is influenced by human, vehicular, and environmental factors that in turn produce damages to people, vehicles, and the environment. Haddon believed that the middle cell of this matrix—crash-phase vehicular modifications—had been particularly neglected in prior thinking by the safety establishment.

As a person, Haddon was described by his admirers as a brilliant thinker whose attitude toward life was serious and determined. His approach to public issues was principled, scientific, and somewhat inflexible. His personal appearance—tall, thin, crew cut, bow tie—reflected his discipline and rigidity. At the same time, Haddon was highly opinionated, articulate, and abrasive. Believing that vehicle packaging was the key to safety progress, he had little interest in state programs or behavioral strategies. He was deeply suspicious of the auto industry's intentions, as was Ralph Nader. Although Haddon and Nader had compatible policy viewpoints, their world views were strikingly different. Haddon saw himself as a scientist, while Nader was a consumer advocate.

Under Haddon's determined leadership, the first two years of

NHSB's existence were extremely busy. By December 1969, twenty-nine motor vehicle safety standards were issued and ninety-five more were proposed, many of them actually codification of industry practice.[68] NHSB also created sixteen highway program standards together with manuals for state safety offices, which were compelled by Haddon to initiate their own highway safety programs.[69]

Haddon's efforts were criticized by all sides. Ralph Nader called the revised standards "virtually meaningless" because they did little to go beyond industry practice.[70] William Stieglitz, a consultant who had helped draft some of the standards, resigned in protest in early 1967 on the grounds that the revised standards issued by Haddon were "totally inadequate."[71] Haddon responded that Stieglitz had proposed "completely unsound" standards that did not satisfy the legal requirement of "practicability."[72] Stieglitz, Haddon charged, wanted to force the industry to build cars that in some ways "approached or exceeded the performance of a Sherman tank."[73] At the same time, the auto industry criticized the Haddon rules on technical and economic grounds and requested that many of them be reconsidered and revised.

Installing Seat Belts

The record shows that the installation of safety belts was not primarily an NHSB accomplishment. The agency simply built on private sector initiatives and responded to political pressures mounted at the state level.

The only companies to offer seat (lap) belts as a factory-installed option as of 1960 were Ford and Studebaker. Other manufacturers, starting in 1956, provided seat belts to dealers for installation on customer request. Postfactory installation was an expensive proposition, especially since seat belt anchors were not offered as standard equipment until the 1962 model year. The anchors were ultimately provided, not due to consumer demand, but in large measure due to political pressure from New York State—especially the efforts of state senator Edward J. Speno.[74]

It was not until January 1, 1964, that front-seat lap belts became standard equipment on all American cars. The auto companies also included rear-seat belts as standard equipment on 1965 models, except for GM, which followed suit on its 1966 models. The provision of seat belts as standard equipment was an anticipatory response to state legislation that began in Illinois and New York in the early 1960s and quickly spread to thirty states by 1966.[75]

Although NHSB played no role in promoting installation of lap belts, it did play the key role in fashioning a requirement that all new cars be equipped with shoulder harnesses.[76] The shoulder harness requirement took effect January 1, 1968, although some manufacturers made the harnesses standard equipment on 1967 models. In the late 1960s the lap and shoulder belts were two distinct systems that required two buckling actions.

NHSB officials were confident that belts would mitigate injuries when worn, but they soon learned that most American motorists did not wear safety belts. The nonuse problem was bad for lap belts (20–40 percent use) and terrible for shoulder harnesses (2–10 percent use).[77] Surprisingly, relatively little was done by Congress or NHSB to encourage belt use—perhaps due to Haddon's belief that behavior could not be modified. For example, the Highway Safety Act of 1966 authorized states to receive federal safety funds if they established a dozen or so specified safety programs. None of those programs addressed belt use. Likewise, no office in NHSB had principal responsibility for promoting belt use. Instead, under Haddon's leadership, agency officials worked on a new rule that would require auto companies to provide "passive" protection in crashes, that is, protection that did not depend on the actions of motorists. Haddon was aware that both Ford and GM had initiated some exploratory work on a new safety system which we now call "air bags."

Conclusion

Passage of the 1966 Safety Act was as much a failure of corporate strategy as a success of the consumer movement. The Big Four had little presence in Washington and little in-house capacity to engage in political behavior. Their responses to the initiatives by congressional entrepreneurs and Nader were consistently too little and too late. The attempt by GM to investigate Nader's private life was a tactical error that reflects both the company's arrogance and its naiveté about public relations.

The decision of the Big Four to send their top executives to testify at congressional hearings is further evidence of their lack of appreciation of how Washington works. The congressional hearing room is designed for the questioners, not the witnesses, and so only in rare circumstances can the witness win the public relations battle with the questioners. As top corporate executives are not trained to perform effectively in this environment, it is not surprising that the public reaction to their performance was skeptical.

What the Big Four should have done was send their appealing, forward-looking, articulate vice-presidents of government affairs— the type of person that the companies had not yet learned they needed.

The most sensible step Detroit took during this period was to cut their losses by turning to a Washington pro, Lloyd Cutler, in search of a quiet compromise. By the time Cutler was hired, the political momentum behind radical legislation was unstoppable. Cutler was able, however, to win some fine points in legislative craftsmanship that served the industry well in subsequent litigation.

Corporate incompetence was, however, not the entire story. The pioneers of crash injury research had provided a reputable technical foundation for a whole new thrust of political activism. Automotive death rates turned upward in the early 1960s, just when Detroit's profits were handsome. The opportunity was seized effectively by congressional entrepreneurs, LBJ, and Ralph Nader. For the media the story was dazzling: life and death, heroes and villains, and intense political competition for recognition from the media. The result was sweeping regulatory legislation and establishment of the National Highway Safety Bureau.

Endnotes

1. Elizabeth B. Drew, "The Politics of Auto Safety," *The Atlantic Monthly* (October 1966), p. 95.
2. Robert I. Field, "Patterns in the Laws on Health Risks," *Journal of Policy Analysis and Management* 1:257–260 (1982).
3. *New York Times* (January 28, 1965), quoted in Drew, p. 95.
4. Joel W. Eastman, *Styling vs. Safety: The American Automobile Industry and the Development of Automobile Safety* (Lanham: University Press of America, 1984), p. 188.
5. Ibid.
6. Fletcher D. Woodward, "Medical Criticisms of Modern Automotive Engineering," *Journal of the American Medical Association* (October 30, 1948), pp. 627–631.
7. Ibid., pp. 629–630.
8. Eastman, *Styling vs. Safety*, p. 191.
9. Ibid., p. 211.
10. Ibid., p. 213.
11. Hugh DeHaven, "Accident Survival—Airplane and Passenger Car," preprint of a paper presented at the SAE Annual Meeting, Hotel Book-Cadillac, Detroit, January 16, 1952, pp. 1–4.
12. Eastman, *Styling vs. Safety*, pp. 232–233.
13. Daniel P. Moynihan, "Epidemic on the Highway," *Reporter* (April 30, 1959).

14. Ibid.
15. Eastman, *Styling vs. Safety*, pp. 159–165.
16. Daniel P. Moynihan, "U.S. Traffic Accident Statistics Useless? Solution of National Tragedy Hindered," *Trial* (June–July 1965), p. 12.
17. Drew, "The Politics of Auto Safety," pp. 97–98.
18. Ralph Nader, *Unsafe at Any Speed: The Designed-In Dangers of the American Automobile* (New York: Grossman Publishers, 1972 [updated]).
19. Ibid.
20. Drew, "The Politics of Auto Safety," p. 96.
21. Sam Peltzman, "The Effects of Automobile Safety Regulation," *Journal of Political Economy* 83:677–725 (1975).
22. William E. Evans and John D. Graham, "Traffic Fatalities and the Business Cycle," *Alcohol, Drugs and Driving: Reviews and Abstracts* 4, no. 1, 1988, pp. 31–38.
23. Eastman, *Styling vs. Safety*, p. 241.
24. Michael Pertschuk, *Revolt Against Regulation: The Rise and Pause of the Consumer Movement* (Berkeley: University of California Press, 1982), p. 21.
25. William Haddon, Jr., Edward A. Suchman, David Klein, *Accident Research: Methods and Approaches* (New York: Harper & Row), 1964.
26. "Congress Acts on Traffic and Auto Safety," *Congressional Quarterly Almanac*, 1966, p. 268.
27. Eastman, *Styling vs. Safety*, p. 245.
28. Ibid.; Drew, "The Politics of Auto Safety," pp. 96–97.
29. Ibid., pp. 245–246; Drew, "The Politics of Auto Safety," pp. 99–100.
30. Ibid., p. 246; Drew, "The Politics of Auto Safety," pp. 99–100.
31. Eastman, *Styling vs. Safety*, p. 245.
32. Pertschuk, *Revolt Against Regulation*, p. 21.
33. Eastman, *Styling vs. Safety*, p. 243.
34. Drew, "The Politics of Auto Safety," p. 97.
35. Ibid., p. 95.
36. Eastman, *Styling vs. Safety*, p. 244.
37. "Traffic, Auto Safety Act," *Congressional Quarterly Almanac*, 1966, p. 274.
38. Pertschuk, *Revolt Against Regulation*, p. 21.
39. "Traffic, Auto Safety Act," p. 277.
40. "Traffic, Auto Safety Act," p. 275.
41. Drew, "The Politics of Auto Safety," p. 97; Eastman, *Styling vs. Safety*, p. 246.
42. Drew, "The Politics of Auto Safety," p. 98.
43. Ibid., p. 97.
44. "Congress Acts on Traffic and Auto Safety," p. 268.
45. Drew, "The Politics of Auto Safety," p. 98.
46. "Traffic, Auto Safety Act," p. 273.
47. Drew, "The Politics of Auto Safety," p. 98.
48. Ibid., p. 95.
49. Ibid., p. 99.
50. Eastman, *Styling vs. Safety*, p. 246.
51. Ibid.
52. Ibid.

53. "Congress Acts on Traffic and Auto Safety," p. 266.
54. Ibid.
55. Ibid.
56. Ibid.
57. Eastman, *Styling vs. Safety*, p. 246.
58. Drew, "The Politics of Auto Safety," p. 96.
59. Ibid., p. 96.
60. Ibid., p. 100.
61. "Highway Safety Act," p. 281.
62. 15 U.S.C. 1381–1481 (Supp. V 1981).
63. Senate Report No. 1301, 89th Congress, 2nd Session, reprinted in *U.S. Code, Congressional and Administrative News*, 1966, pp. 2701–2711.
64. Ibid., p. 2712.
65. 15 U.S.C. 1381–1481 (Supp. V 1981).
66. Haddon, Suchman, and Klein, *Accident Research*.
67. "Haddon Matrix: Eliminating the Mumbo Jumbo," *Status Report: Highway Loss Reduction*, Insurance Institute for Highway Safety (Sept. 9, 1986), p. 8.
68. "Auto Safety," *Congress and the Nation, Vol. II*, Congressional Quarterly, 1970, p. 804.
69. Ibid.
70. Ibid.
71. Ibid.
72. Ibid.
73. Ibid.
74. Lawrence J. White, *The Automobile Industry Since 1945* (Cambridge: Harvard University Press, 1971), p. 241.
75. Ibid.
76. Ibid., p. 244.
77. Ann Grimm, "Use of Restraint Systems: A Review of the Literature," *The HSRI Review* 11:11–28; 1980; "Auto Safety: Passive Systems Pushed," *Steel* (20 October 1969), pp. 7–8.

Chapter 3

THE "TECHNOLOGY-FORCING" STRATEGY

A contemporary reader of the 1966 Safety Act and its legislative history might be inclined to predict that implementation of the legislation would prove to be a war between the administrative agency and the auto industry. As the agency issued performance standards challenging the technological capacities of the manufacturers, firms in the industry would draw together to contest feasibility and launch a political counterattack. This image proved to be an erroneous—or at least simplistic and premature—version of how the story of occupant-protection regulation would actually unfold.

The so-called "technology-forcing" approach embodied in the Safety Act instigated some complex responses from corporate strategists. Suppliers to the auto industry took stances in favor of or opposed to ambitious rules on the basis of what such rules would do for their commercial position. Likewise, individual car manufacturers saw technology-forcing rules as a tool in their competition with rival manufacturers. There was no campaign by "the industry" to "capture" the regulators, but instead a campaign of particular suppliers and manufacturers to manipulate the regulatory outcome in their favor. As we shall see, firms behaved strategically and were not reluctant to support regulation if it might strengthen their relative competitive position.

An Innovative Idea

The pioneer developer of the automobile air bag was Eaton, Yale and Town, Inc., a Cleveland-based supplier of the automotive industry. In the 1960s Eaton developed its own research center where engineers were given a broad mandate to generate ideas for new automotive products. The air bag was one of their promising ideas, with Ford Motor Company enlisted as Eaton's first major customer.

In 1968 engineers from Eaton and Ford presented a paper to the Society of Automotive Engineers in Detroit that described how an automobile air bag would work in a crash.[1] By the end of 1968 Eaton had spent more than $1 million designing and developing the first experimental air bag system ("Auto-Ceptor") for Ford at the supplier's facilities in Southfield, Michigan.[2]

The approach taken by Eaton engineers was simple in concept. Crash sensors in the structure of the car were set to trigger deployment mechanisms within the passenger compartment whenever deceleration forces in the forward direction exceeded five G's (five times the force of gravity). The deployment mechanisms would inflate large nylon cushions (stored within the steering wheel and instrument panel) into the car interior within 0.04 second—before the car occupants begin to move forward. During the "second collision," occupants would strike the inflated nylon cushions instead of the steering wheel, dashboard, or windshield. Holes in the ends of the bag would cause the cushions to deflate rapidly after crash forces were absorbed.

In extensive laboratory crash testing, Eaton engineers demonstrated that the air bag system could provide excellent crash protection to occupants in frontal barrier crashes. In one set of tests, baboons protected by air cushions survived 50 percent more crash forces than baboons restrained with conventional lap belts and shoulder harnesses.[3] The cushions appeared to be especially effective in providing head, facial, and abdominal protection in high-speed crashes in which belts inflicted injuries and failed to restrain adequately the forward motion of occupants.

The Auto-Ceptor system was designed to supplement lap belts and replace shoulder harnesses. Lap belts were still needed to help position the occupants for air bag deployment and to provide occupant protection in side collisions and rollovers—crash modes in which air bags were not designed to inflate. In the long run, Eaton engineers hoped to add inflatable cushions to the side and top of the car to provide occupant protection in nonfrontal collisions.[4]

Technical critics of the Eaton design believed the 5 G deployment threshold was too lenient; the air bag would deploy in many minor accidents when injuries would not be expected. Further, replacement of the bags would be expensive, and the explosive force of the air bag would cause injuries to occupants in some minor crashes where no injuries would otherwise have been expected.[5] As we shall see, the Eaton design was indeed somewhat primitive compared to the advanced air bag systems that are being marketed in the late 1980s.

Interest in Washington

Eaton's work on air bags was monitored with a sympathetic eye by the first director of the National Highway Safety Bureau (NHSB), Dr. William Haddon, Jr. Haddon saw improved "passive" protection as the agency's number one priority. Haddon wanted to accelerate the pace of air bag development, but he chose not to invest government dollars in research and development. As a result of a provision in the 1966 Safety Act, the inventors of air bags might be unable to collect royalties if their research and development efforts were funded directly by taxpayers.

In August 1968 Haddon organized a "state of the bag" meeting among Eaton's key engineers, representatives of NHSB, and the "Big Four" domestic automobile manufacturers. At this early meeting, the unresolved problems with the technology were uncertain long-term reliability, potential hearing damage from deployment noise, danger to out-of-position occupants, and the possibility of inadvertent activation of the systems.[6] Despite these unresolved concerns, Ford's chief car engineer, Robert B. Alexander, predicted at the meeting that Auto-Ceptor would be ready for limited real-world application within two years.[7] Eaton's engineers were more ambitious, suggesting that the system should be a standard item on 1971 cars.[8]

In the spring of 1969 the air bag received national attention at public hearings before Senator Warren Magnuson's Commerce Committee. The acting director of NHSB, Dr. Robert Brenner, told the committee that the air bag might prove to be the key to safety for small-car occupants. Brenner, who had moved up from the deputy directorship, was a temporary replacement for Haddon, who had been asked to resign by John Volpe, the new secretary of transportation in the first Nixon administration. At the Senate hearings, Ralph Nader called the air bag "an exciting development" but deplored the relatively small amount of investment in research and development. Eaton engineers declared that problems with the device had been solved during years of research and testing.[9]

Engineers at General Motors felt that the claims of Eaton's engineers were preposterous. In their own testing of air bags, GM had inadvertently killed a baboon and injured others due to excessively aggressive air bag deployment forces. David Martin of GM recalls that the Eaton position was "crazy" at the time because no one had yet done the technical work to assess the injury risk from inflation forces—a necessary step to assure a safe product and measure compliance with any standard in "objective" terms (as was required by the 1966 Safety Act).[10]

Prototypes of Eaton's Auto-Ceptor were supposed to be produced that summer. According to Dr. Peter Haas, Eaton's marketing manager for Auto-Ceptor, "our job now is to carry [the air bag] to the marketplace."[11] Though many safety engineers employed by the Big Four considered the air bag to be a promising idea, there was skepticism about whether it was ready for installation into new cars in the near future. Even at Ford, where development work was progressing rapidly, the internal skeptics were citing reliability and cost as "hang-ups" that precluded immediate installation of air bags into commercial vehicles.[12]

Nader's New Center for Auto Safety

After passage of the 1966 Safety Acts, Ralph Nader built a small organization of activists who monitored NHSB and pressured the industry on auto safety issues. In 1969 Nader recruited several young attorneys at the Harvard Law School to join his cause.

One of Nader's recruits was Lowell Dodge, who became a Ralph Nader Auto Safety Fellow for Consumer's Union. Dodge ultimately became the founding director of the Center for Auto Safety, an organization funded by Nader and Consumer's Union. Dodge recalls that the passive-restraint standard was "our first priority" because of its "tremendous lifesaving potential."[13] The standard was also viewed by Nader and Dodge as a vehicle to pressure Detroit into developing and marketing new approaches to vehicle crashworthiness.

Dodge developed close contacts with those members of Congress and their staff who supported the auto safety movement. The key staff people at the Senate Commerce Committee were Mike Pertschuk and Lynn Sudcliffe because they had access to Senator Warren Magnuson. Senators Abraham Ribicoff and Gaylord Nelson were also supportive of Nader's agenda, although Nelson's interests were primarily in tire safety. As a result of these contacts, NHSB officials—especially chief counsel Larry Schneider—were cooperative with Dodge and Nader.

Enter John Volpe

Safety engineers at NHSB saw the air bag as an excellent opportunity to fulfill their "technology-forcing" role in car safety. Although Ford had played an instrumental role in air bag development, the supposition among NHSB officials was that vehicle manufacturers, if left to their own devices, would be slow in transforming the air bag from the developmental stage into full-scale production. The charge of indifference about safety was a central feature of Nader's case against corporate management and was partly responsible for passage of the 1966 Safety Act.

Air bag advocates found an unexpected ally in John Volpe, President Nixon's nominee for the cabinet position of secretary of transportation. Volpe was a former governor of Massachusetts whose political roots were in the moderate-liberal wing of the Republican party. While Volpe was known as a business-oriented Republican (his family acquired wealth through the Volpe Construction Company), he was also a devout Catholic who was described even by his critics as a "decent, kind, and compassionate man."[14] From the beginning of his tenure at DOT, Volpe made clear that he was eager to carry out the lifesaving vision of the architects of the 1966 Safety Acts. He saw NHSB as a home for "people-oriented" programs.

Although originally skeptical about the air bag, Volpe quickly recognized the lifesaving potential and political sex appeal of this imaginative device. After seeing the human side of air bag protection portrayed in crash films by NHSB staff, Volpe became dedicated to the large-scale use of the technology. According to one NHTSA engineer, "Volpe's attitude toward the air bag became so tremendously supportive that he was an inspiration to air bag engineers both in NHSB and in Detroit."[15]

In July of 1969 Volpe approved an advanced notice of proposed rule making from NHSB, calling for "inflatable restraints" to be installed in all new cars, trucks, and buses as soon as possible. The notice contained no specific deadline, but Volpe's public statements made clear that he was thinking about air bags on some 1971 models and all 1972 models.[16] Volpe called for public hearings in late August where he would receive comments on the proposal.

Doubts in Detroit

In Detroit Volpe's proposal was the subject of much grumbling and criticism. *Business Week* quoted the immediate reactions of one auto industry executive: "It's insanity to require something as unproven as this as early as the 1972 model." David Martin of GM recalls that this was also the reaction of GM's best safety engineers.[17] Other industry analysts expressed doubts about whether consumers would be willing to foot the "$200 tab" for the most expensive safety device in the history of the industry.[18]

Just before the Volpe hearings began, Ford engineers revealed "tentative plans" to install a front-seat, passenger-side air bag system in the 1971 Mercury Marquis.[19] At a successful media demonstration of the air bag, there were reports that the new Ford air bag seemed to be less noisy and startling than the original Eaton design. In a separate development, Cornell Aeronautical Labs of Buffalo—a major auto safety research center—reported that the performance of Eaton's air bag system in recent crash tests had been "very impressive."[20]

At the Volpe hearings in late August, the testimony of the automakers conveyed a common message: The air bag concept was promising, but more development and testing were needed. Support for the concept was not restricted to Ford; officials from General Motors were also enthusiastic. In fact, GM had quietly launched its own air bag program, extending some earlier work begun in the 1950s. Foreign manufacturers were behind in air bag development but not necessarily opposed to the idea. Volkswagen's resident engineer in the United States, Henning Kirstein, also told Volpe that "air bags should be incorporated into vehicles as soon as those technical difficulties involving reliability, maintenance, and serviceability have been overcome."[21]

Within the auto industry there were many skeptics of the air bag, but few were willing to be vocal on the subject. An exception was Roy Haeusler, Chrysler Corporation's chief safety engineer. Haeusler was one of the earliest and strongest advocates of seat belts in the entire auto industry. He felt that it was important that people participate in safety. Haeusler himself was a religious belt user who even wore crash helmets on occasion when driving cars. He was especially frightened by air bag advocates who wanted to remove lap belts from cars and replace them with various types of passive protection.

Haeusler repeatedly told reporters that government and industry were concentrating too much effort on the air bag, which, he argued, was five to six years away from even experimental use.[22] Other safety ideas, such as stronger seats and more comfortable belt systems, were being neglected. Contrary to popular thinking at the time, he saw no chance that air bags would be installed in some 1971 models. At the Volpe hearings in late August, Haeusler emphasized the need for an industry-government program to study air bag hazards and product liability risks that might be created by air bags. These steps should be taken, he argued, before any decision was made about a mandatory standard.[23]

Officials from Chrysler, Volkswagen, and Mercedez-Benz testified before Volpe that their companies could not meet a 1972 deadline. Ford and GM officials were silent on this topic at the public hearing. Ford planned to dominate media coverage at the hearing with a live demonstration of how an air bag works. When Ford's chief body engineer, Stuart Frey, pressed an electric switch to trigger the bags in an empty red Galaxy, nothing happened. Accusations of intentional failure were made by consumer advocates but never proven. The embarrassing moment received national media coverage, perhaps foreshadowing the forthcoming delay in Volpe's rule making.[24]

At Eaton the future nonetheless seemed bright. The *New York Times* reported in late August 1969 that Eaton's stock "was one of a small handful posting new 1969 highs on the New York Stock Exchange."[25] The air bag was described as "a chief reason for the current popularity of Eaton, Yale and Town, Inc. among investors."[26]

In Detroit there were complaints that Eaton was "looking for a bonanza" from royalties in a mandated monopoly market.[27] Just weeks prior to Volpe's advanced rule-making notice, patents for the Auto-Ceptor system were granted to William R. Carey (project engineer) and David P. Haas (marketing director) of Eaton, Yale and Town, Inc.[28] In October 1969 Eaton acquired a pilot production and engineering facility and was expecting to produce air bag systems for testing by companies in early 1970. Following Volpe's rule-making action, Eaton's list of clients expanded rapidly to include all of the Big Four as well as several foreign companies.

An Acceptable Republican

Immediately following the August hearings, Volpe's attention was consumed by the need to appoint new leadership at NHSB. This task proved to be difficult. Nader and Haddon, fearing that Nixon and Volpe would appoint a pure politician to run NHSB, raised the visibility of the NHSB appointment through their contacts on Capitol Hill and made clear their intention to fight an unacceptable nominee.

Several prospective appointees told Volpe that they were not interested in the job because of its subordination to the Federal Highway Administration (FHWA) in the DOT bureaucracy. Other prospects were rejected because they were considered to be "too close" to the auto industry. Meanwhile NHSB was being run by Robert Brenner, who critics said was a "poor people manager."[29]

In the face of these problems, Volpe decided to take a more fundamental look at the status of NHSB in the DOT hierarchy. Haddon and Nader had been arguing for years that NHSB should be given independent status from FHWA, where highway construction interests were said to be dominant. An alternative view was that NHSB was a recalcitrant bureau that needed the supervision provided by FHWA. According to a career NHSB official, "FHWA Director Frank Turner was very receptive to our initiatives on motor vehicle safety regulation and was quite willing to go to Congress for money on our behalf."[30]

For recommendations on this issue Volpe turned to Douglas Toms, the director of motor vehicles for the state of Washington. As a consultant to Volpe, Toms argued that the NHSB's status within DOT needed to be upgraded. The current arrangement was, Toms argued, hampering the recruitment of competent safety professionals at NHSB. Moreover, the NHSB director's lack of access to the DOT secretary was weakening the voice of safety advocates in policy discussions.[31]

Volpe received the Toms report with praise and responded by offering Toms the directorship of NHSB. Toms accepted on the condition that the bureau be removed from FHWA and placed directly under the DOT secretary. Volpe agreed.

As Toms describes the situation, "I was probably more acceptable to the safety zealots than I was to the White House."[32] Toms had doctoral training in traffic safety at the University of Michigan and state experience as a manager of a motor vehicle agency. During his earlier years of academic research on traffic safety, he had developed his own reputation in the field and had earned the professional respect of Haddon, who was one of the scientific visionaries in the field.

Toms was also a lifelong Republican who had been active in politics in the Seattle area. He had been appointed to his government post by Republican Governor Dan Evans and had indirect links to John Ehrlichman and Bud Krogh, Seattle attorneys who went to key jobs in the Nixon White House. The two senators from Washington, Warren Magnuson and Henry ("Scoop") Jackson (both senior and influential Democrats), let the White House know about their support for Toms.[33] Magnuson later chaired Toms's confirmation hearings in the Senate, which proved to be noncontroversial.

The appointment of Toms was significant in two respects. By looking to a safety professional to lead NHSB, Volpe expressed his commitment to the agency's safety mission. But Toms was quite different from his predecessor at NHSB, William Haddon. The NHSB's early years were stormy in part due to Haddon's adversarial leadership style. In contrast, Toms was considered a "natural born diplomat."[34] His gregarious personality and positive attitude toward the automobile were expected to reduce Detroit's suspicion of NHSB. According to Haddon's aide Joan Claybrook: "Toms was a friendly and charming man with a taste for the good life. He had a sense of bravado and a gutsy attitude toward Detroit."[35]

The industry soon learned that Toms was also very committed to the air bag. According to a career NHSB official, Toms identified "three priority areas" when he came to the agency: "air bags, alcohol, and the Experimental Safety Vehicle."[36] In the spring of 1970, his first formal move on the air bag issue was a notice of proposed rule making that called for "passive restraints" to be installed in all new cars beginning January 1, 1973.[37]

Naturally, Toms's proposal was music to the ears of air bag suppliers. Along with Eaton, Yale and Town, Inc., the Energy Systems Division of Olin Corporation in East Alton, Illinois (which was working on inflation systems) envisioned a surge in their order books.[38] Dozens of other companies were positioning themselves to be suppliers of the various air bag components. The prices to be charged by the suppliers were a matter of considerable uncertainty and of great concern among the Big Four.[39]

Despite Toms's support for the air bag, some of his early actions as head of NHSB were controversial. According to Toms:[40]

Haddon had hired a blend of good people and gadflies. There were several disgruntled auto engineers who wanted to get even with Detroit. There was Joan Claybrook, who had been Haddon's administrative assistant and had close ties to Nader. Joan and I knew when I came that it would be best for her to leave. When she resigned to join Nader, my relationship with Senator Magnuson's committee (especially staff member Mike Pertschuk) became a little strained.

A career NHTSA official recalls that Toms had "no hesitation about firing people" whom he regarded as incompetent.[41]

Ed Cole's Strategy

Daniel Patrick Moynihan was a strong advocate of the 1966 Safety Acts. Later, in his role as a White House consultant to President Nixon, he helped Volpe and Toms organize NATO's First International Conference on Passive Restraints in the summer of 1970. The affair was cosponsored by the domestic auto industry and the

U.S. Department of Transportation. It was held at GM's Proving Grounds in Milford, Michigan, and was supposed to be a forum for "technology exchange" between domestic and foreign producers. DOT officials invited reporters from throughout the world, hoping to publicize the theme of Secretary Volpe's opening address: "Passive restraints are systems whose time has come." It was expected that automotive engineers would use the occasion to reveal their latest accomplishments in air bag development.[42]

What transpired was somewhat unexpected. A groundswell of opposition to the air bag in general and the NHSB proposal in particular had arisen in Detroit. The opposition was strongest at Chrysler, Ford, and American Motors. The leadership at GM remained positive about the air bag but was critical of NHSB's January 1973 deadline.

The president of General Motors Corporation, Ed Cole, was a creative engineer in his own right and a strong proponent of the air bag. He saw air bags—in contrast to belts—as a way to make driving both safer and more enjoyable. Actually, GM had been working on the air bag concept since the mid-1950s. Cole himself was known to hold a patent on air bag technology.[43] After assuming the presidency of GM in 1968, Cole initiated the massive commitment of resources, effort, and brainpower that promised to bring the air bag to the marketplace in the 1970s.[44]

At the NATO Conference, it became apparent that GM had surpassed Eaton as the world's leader in air bag development. In particular, Robert McLean, project manager for the safety cushion at GM's Fisher Body Division, "seemed to know more about the air bag than anyone else at the Conference."[45]

Cole was also receptive to Toms's personal approach to air bag promotion. Toms would make frequent trips to Detroit and Europe to chat with auto executives and infect them with his enthusiasm for both the air bag and the Experimental Safety Vehicle (an experimental car with remarkable safety performance).[46] Claybrook recalls with a smile: "Nader and I knew that Toms had become pretty chummy with Ed Cole. We didn't object, of course, since both men were pro air bag!"[47]

Cole and Toms also had their differences. Despite his enthusiasm for the air bag, Cole made clear at the NATO Conference that the NHSB's January 1973 deadline was unrealistic. It represented a time frame that GM had never met previously in the introduction of major design changes.[48]

Cole went further and disputed openly with DOT officials about the wisdom of another new NHSB regulation that required combination lap/shoulder belts, instead of separable lap and shoulder belts, to be installed in all new cars beginning January 1, 1972. Engineers at Ford had designed an integral three-point belt system that impressed NHTSA's engineers. GM officials were not impressed. In Cole's words: "It seems to me that you're taking money and people away from the air bag program, thereby delaying our ability to solve the many problems [with air bags that are] remaining."[49] Cole was of the opinion that air bags would ultimately be a replacement for, rather than supplement to, belt systems. On this issue Cole was considered stubborn by some of GM's working-level safety engineers.[50]

Ford and Chrysler Resist

The NATO conference also signaled a shift in Ford Motor Company's posture toward air bags. Ford supported Eaton's early R&D efforts, but the company had a strong internal group of air bag critics. Executive Vice President Lee Iacocca, for example, opposed air bags and told reporters "it scares me." Just before the NATO conference, Ford dropped plans to install the air bag on some 1971 models. Stuart M. Frey, Ford's chief body engineer, pointed to several flaws in the Eaton system that advised against near-term marketing. There was possible danger to standing children, excessive noise, liability risk in cases of failure to deploy or inadvertent deployment, and uncertain system longevity. The theme from Ford's representatives at the conference was that the air bag needed more development and testing before it could be introduced on a large-scale basis.[51]

This theme was echoed by Sydney L. Terry, vice president of Chrysler Corporation. As the "weak sister" in the industry, Chrysler was viewed by NHSB officials as a foot-dragger on air bags.[52] After learning that NHSB was contemplating an $11 million real-world air bag testing program to be completed by 1976, Terry commented: "It's a shame that we'll have to prostitute our cars by installing these bags in 1973, and to make guinea pigs out of our customers before the final results [of such tests] are in."[53] All of the manufacturers were apparently very concerned about the liability risks posed by air bags. One industry journalist opined that "the [liability] concern is of such magnitude that the companies are likely to fight this safety standard as they've never fought any other."[54]

The NATO Conference's final address by NHSB Administrator Toms indicated that the agency was determined to move ahead, despite the industrial opposition. He made four major points:[55]

1. The government sought to eliminate manual belts (which were rarely used) in favor of "passive restraints."
2. The government could assume liability risks for the manufacturers, although this would require new legislation.
3. Mandatory seat belt wearing laws were not considered a politically feasible alternative to air bags.
4. An optional air bag program by manufacturers would not be a satisfactory substitute for regulation, even as a phase-in strategy.

Toms emphasized that "our objective is to cause full air bag installation to happen as fast as we can humanly do it."[56]

After the conference, Secretary Volpe reaffirmed NHSB's position on air bags in a public statement of the Nixon administration's approach to automobile safety. He cited the air bag as the nation's number one transportation safety priority.[57] In response to criticism of the proposed January 1, 1973, deadline, Volpe said "somebody's got to prove to me it cannot be done."[58]

Cole's Phase-In Plan

During the public comment period on NHSB's proposal, reactions from Detroit were quite critical. The exception was General Motors, which offered a concrete alternative to NHSB's plan. GM acknowledged that air bags could be installed in all new cars by the fall of 1972 and in all new trucks by the fall of 1974, but stressed that such an ambitious timetable was not in the public interest. GM had not yet installed an actual production-type air bag in a passenger car and operated it on public roads. Indeed, no one had any real-world experience with the system. As an alternative, GM President Ed Cole advocated a four-phase plan that—while contingent upon resolution of lingering technical questions—called for:[59]

- Immediate development and production tooling for air bags.
- Production of 25,000 prototype air bag components by January 1972 to be used for exhaustive laboratory and road testing
- Production of 150,000 cars with air bags for sale to consumers by the fall of 1972.
- Availability of air bags as (1) an option on some 1973 models, (2) standard equipment on one million cars beginning January 1974, and (3) an option on all remaining 1974 passenger cars.

GM argued that the phase-in period would allow for both better quality control in production and a less costly return to manual belts if serious problems with the air bag developed during real-world testing.

According to Toms, some agency officials saw considerable merit in GM's proposal. Cole had explained to Toms on several occasions that a phase-in period for the air bag was critical to avoid production defects and to build confidence in the technology.[60] One NHTSA engineer believes that this is where Volpe was "too supportive of the air bag."[61] In conversations with senior agency officials, Volpe would insist that "those auto companies are so big and powerful that they can do anything they want."[62]

Volpe and Toms were also under pressure from Capitol Hill to proceed with their May 1970 proposal that called for passive restraints in all new vehicles sold after January 1, 1973. For example, a June 23, 1970, letter to Volpe and Toms from sixty-two congressmen conveyed "strong congressional support" for the May 1970 proposal. The letter emphasized that "any further delay in the effective date of the air bag requirement is without justification. You should stand firmly by your current timetable."[63] On the Senate side, Senator Warren Magnuson's staff—especially Mike Pertschuk—were watching NHSB's actions on air bags with a critical eye. Volpe and Toms saw a passive-restraint mandate as more appealing than a phase-in and therefore decided to "go for it all."[64]

R&D Flourishes

During the beginning of NHSB's rule making on occupant restraints, the phrase "passive restraint" was understood to mean either air bags or other (as yet) undeveloped safety systems that would not require action by vehicle occupants. In 1970 air bags were the only known viable alternative, but as opposition to air bags increased, investment in development of alternatives also increased.

In the fall of 1970 Hamill Manufacturing Company, a major producer of seat belts, finished a two-year $250,000 "safety-blanket" research program.[65] Working on the rear seat, Hamill engineers envisioned a blanket stored in a compartment at the lower edge of the back of the front seat. When crash forces were detected, a propellant inside the trunk would explode, causing a heavy belt in the top of the blanket to instantly draw tightly against the rear seat at chest level. The blanket would be made of energy-absorbing fabric and would be less expensive than the air bag.

Allied Chemical Corporation also made large investments in alternative restraint systems. One of its more innovative ideas was an inflatable belt system that would do a better job than conventional belts of absorbing and distributing crash forces. The draw-backs of inflatable belts were that they were somewhat cumber-some and uncomfortable.[66]

Other companies, such as Volkswagen, were looking to develop "passive" safety belt designs. Volkswagen's look at alternatives to the air bag was one component of a larger effort to develop a safe 2,000-pound car. VW sought to squelch complaints that small cars were necessarily "motorized death traps."[67] Under the creative leadership of Ulrich Seiffert, VW designed an automatic shoulder belt with knee pads that could satisfy NHSB's frontal crash perfor-mance criteria. This discovery was so surprising to NHSB engi-neers that it was widely disbelieved.

Tough "Final" Rule

When the National Highway Traffic Safety Administration (NHTSA)—NHSB's new name after executive reorganization—issued its final rule in November 1970, air bags were not considered the only means of compliance with the performance standard. The rule stated only that "passive" protection must be provided for front-seat occupants of new cars by July 1, 1973, and for all seating posi-tions by July 1, 1974. Administrator Toms stressed that non–air bag alternatives—including fixed cushioning of vehicle interiors, self-fastening seat belt systems, and crash-deployed blankets or nets—"are equally acceptable methods to the extent they satisfy the perfor-mance standard in crashes."[68]

The July 1973 deadline in the final rule represented only a six-month extension from the January 1973 deadline in the proposed rule. Toms wanted a tight deadline. He stressed that it would "not be in the public interest to introduce [passive protection] at the pace preferred by the slowest."[69] The agency did issue a longer (18-month) delay of the rear-seat protection requirements on the grounds that much more engineering work was required to protect occupants in these seating positions.

While NHTSA recognized that alternatives to air bags were possi-ble compliance strategies, the agency was still committed to (and arguably enamored with) the air bag concept. The tight NHTSA deadlines irritated air bag critics, who saw them as a ploy to pre-clude innovative alternatives to air bags, which might require more research and development time.

NHTSA also began to invest in air bag research. For example, in late 1970 NHTSA made a $500,000 research grant to the Cornell Aeronautical Labs to investigate the feasibility of certain innovative air bag concepts.[70] One idea was to devise a compartmental air bag within an outer shell that would provide protection in multiple impact crashes. Another idea was to design systems with several air bags instead of one so that bags could be inflated "around the occupant" to provide protection in oblique and side impacts. Oddly shaped air bags might also be employed to prevent head-knocking and damage from side impacts.

At one point Secretary Volpe publicly expressed interest in the notion of inflatable "head bags" instead of head restraints on the grounds that fixed head restraints interfere with a driver's vision.[71] Olin Corporation, the inventor of this idea, saw the head bags as a promising way to further expand the air bag market. Many of these novel ideas never proved to be practical.

The government's pro–air bag position received crucial technical support from numerous career people at NHTSA. One of the most important contributors was Robert Carter, an aerospace scientist recruited by Haddon who initially ran NHSB's program on crash-worthiness research and later ran the agency's program on vehicle rule making and enforcement. Carter's philosophy was that big advances in car safety required "revolutionary change, not evolu-tionary change."[72] He was described by several of his contemporar-ies as the driving force behind air bag research and development at NHSB during the early years.[73]

Detroit Goes to Work

The challenge of designing a safe yet effective air bag system was formidable. Although NHTSA's rule required only that satisfactory 30 mph crash tests be achieved for a 50th percentile male dummy, auto engineers knew that they needed to go beyond the terms of the rule to make the device commercially acceptable. No responsi-ble manufacturer could install a device that would only work at one speed or for only one size of occupant. More development needed to be done to prevent bag-induced injury of five-foot females seated close to the bag and to protect six-foot-plus males seated far from the bag. Engineers were also worried that a system optimized for the 30 mph barrier crash test specified by the government might be stiffer than desirable at lesser speeds and might actually cause injuries in minor crashes.

NHTSA's aggressive regulatory program caused manufacturers to respond by making major new investments in passive-restraint development. It was becoming increasingly clear to the auto compa-

nies that Washington was serious and any company that fell behind
might be placed at a competitive disadvantage.

Engineers at General Motors worked on improved air bag systems and alternative compliance strategies. Leaders of this crash
effort were Robert McLean and David Campbell (air bag design)
and Trevor Jones (sensors and electronics). New GM air bags were
deployed with a 70 percent reduction in noise levels relative to
earlier designs.[74] Tests with human volunteers revealed no hearing
damage.[75] To lessen the chance of injury to children, GM engineers redesigned the air bags so that deployment occurred upward
rather than directly rearward. They also found a way to prevent a
car's doors from springing open when the bags are inflated.[76]

Meanwhile, GM also investigated the injury-mitigation potential
of new windshields, extra interior padding, and modified energy-
absorbing steering columns. This line of inquiry was so promising
that President Cole decided to scrap his plan to offer air bags on
some 1972 models.[77] While Cole preferred the air bag concept, he
stated publicly that the business and liability risks were much
lower for the interior-improvement approaches.[78]

Ford Motor Company also worked on several compliance strategies. It continued its air bag development work, including the
building of fifty Mercury cars to test all phases of the system except
bag deployment. Ford engineers were especially interested in the
combination of lap belts and air bags because the lap belts would
help control the position of occupants during frontal crashes and
restrain their movement during rollovers.

Resistance to the air bag at Ford was considerable, especially at
the highest levels of corporate management. Chairman Henry
Ford II, who was apparently embarrassed by several public mis-
fires of Ford's air bag, opined on the "Today Show" that the air bag
was nothing but "baloney."[79] The result was that Ford's prime
compliance alternative became an "ignition-interlock" system. The
idea was to route the ignition wiring through seat belt buckles, thus
forcing motorists to buckle safety belts before the car could be
started. At several meetings with Volpe and Toms, Henry Ford II
made a personal appeal that the interlock was a better idea than the
air bag.[80]

Roy Haeusler and his colleagues at Chrysler also argued that the
air bag was not the answer. Their reasons included danger to out-
of-position occupants, lack of occupant protection in nonfrontal and
multiple-impact crashes, unreliable sensor performance, excessive
noise levels, inadvertent activations, and enormous tort liability
risks. Chrysler had different ideas.

For front-seat occupants, Chrysler was planning to install a system of interior padding and "semi-passive" lap belts in 1974 models. Instead of rear-seat air bags, Chrysler engineers contemplated crushable containers filled with energy-absorbing plastic foam or thin-walled metal tubes. These would be strategically placed in a large cushion in front of rear-seat passengers. Unlike the air bag, the containers would remain intact to provide occupant protection in multiple-impact collisions.[81] According to Roy Haeusler, these ideas were "dreams" that never proved to be practical—except for the interior padding schemes.

In Europe, Toms helped stimulate substantial interest in both the air bag and the Experimental Safety Vehicle. Daimler-Benz emerged as the safety leader among European manufacturers. According to Toms, "Karl Wilfert of Daimler-Benz knew more about the air bag than anyone outside the United States."[82] Volvo also became interested in air bag development, although to a lesser degree than Daimler-Benz.

Suppliers Compete

During the months after Volpe's ruling, Eaton found itself in a very delicate position. After four years and $10 million of research and development on air bags, the leadership of Eaton was eager to market the device. The company was spending at a rate of $4 million annually and had a production capacity of 50,000 air bags per year.[83] The difficulty was that some auto executives were convinced that Eaton had sold Ralph Nader and NHSB on the air bag before they had sold their prospective buyers. Eaton's Chairman, E.M. DeWindt, acknowledged that the car manufacturers were sensitive about air bags because they were in effect being mandated by government rather than being developed normally through industrial channels.[84]

Toms recalls that Eaton was not the most aggressive firm at NHSB from the perspective of lobbying. Eaton's primary competitors in the air bag business were Allied Chemical Corporation, Fairchild Industries, and Talley Industries. Allied Chemical, in particular, had hired several full-time lobbyists who spent much of their time presenting their firm's case to officials at NHTSA and to staff on Capitol Hill.[85] DeWindt knew he was in a competitive environment and devoted most of the firm's energies to fostering the relationship with Ford.

DeWindt was also worried that some of his competitors would persuade manufacturers to go with ignition-interlock systems instead of air bags. Charles H. Pulley, president of the American Seat

Belt Council (a trade group of seat belt makers), was touting the ignition-interlock system with some success. For instance, the system was tested by Ford President Lee Iacocca in his personal car.[86]

Toms and the other air bag advocates viewed Pulley as a "troublemaker."[87] Like Haeusler at Chrysler, Pulley was a seat belt fundamentalist. But unlike Haeusler, Pulley was primarily an entrepreneur and an executive at Irvin Industries (a firm that produced belt systems). Pulley was concerned about the future of his business because NHTSA and GM engineers were planning to give up on active belt systems entirely.[88]

In these early years a discernible tension betweeen seat belt and air bag advocates emerged, characterized by mutual distrust and commercial motives. A generational factor may also have been operating. Air bag engineers and their managerial allies (e.g., John DeLorean at GM) tended to be young, brash, smart, arrogant, while seat belt promoters were older and more conservative, though no less vocal.

Unpopular Compromise

Although work on various safety systems progressed rapidly, Toms came to the conclusion in March 1971—in partial response to requests of automobile manufacturers—that the industry would need more time to comply with certain aspects of the rule. A revised rule was issued that provided manufacturers a one-month extension in the deadline for front-seat passive protection in frontal crashes.[89] This change in the deadline (July 1, 1973 to August 1, 1973) was adopted to coincide with the transition in models.

More importantly, the passive-protection requirements for nonfrontal crashes and rear-seating positions were delayed from July 1, 1974, to August 15, 1975.[90] The result was a two-year phase-in of comprehensive passive protection. In the interim, cars would be required to have manual lap belts with warning devices that signal both audibly and visually when the occupants of front outboard seats were unbelted.

The new rule drew much criticism. Lowell Dodge of the Center for Auto Safety recalls that such delays were "traumatic" among auto safety activists.[91] Ralph Nader demanded that Secretary Volpe give a "detailed justification" of the delays granted and he saw this change as a "clear example of selecting the lowest common denominator for government regulation."[92]

Industrial officials were also unhappy with the new rule. In Detroit there was much talk about court action being the only alternative to the air bag. The idea was to go to court and ask for relief from the standards on the grounds that they are neither practicable nor feasible within current technical knowledge.[93]

Conclusion

The 1969–1971 period illustrates how corporate strategy needs to change in response to technological innovation and a new regulatory environment. The car manufacturers and their suppliers were not oblivious to this principle.

Although Eaton and Ford got a head start on the air bag, GM's Ed Cole knew that he could deploy GM's rich technical resources and beat Ford in air bag development. When he did so, Ford switched course and pursued alternatives to the air bag. Volkswagen, meanwhile, ignored the air bag battle and invested R&D in a creative automatic belt system, the kind of device that air bag zealots at NHTSA believed was impossible to develop. As we shall see, this discovery by VW would have a profound impact on the entire issue for the next fifteen years. Chrysler, the industry's weak sister, struggled in vain to find a corporate strategy. It ultimately realized that from its point of view, the best outcome was no passive-restraint regulation.

For suppliers to Detroit, the new regulatory environment raised some interesting strategic questions. Is a passive-restraint rule sufficiently certain to justify substantial R&D investments in new technology? Eaton and others said yes. Should a supplier spend time selling its technology in Washington, or should it concentrate on building its client base in Detroit, Europe, and Japan? Eaton and others did both. Seat belt makers, more than anyone else, had much to lose from passive-restraint regulation. Should they fight the regulation, or launch a crash R&D program into non–air bag technologies? Irvin Industries and others did the former vocally and the latter quietly. In the chapters ahead, we will learn which corporate strategies proved to be the most successful.

The 1971 passive-restraint rule was characteristic of the Volpe-Toms administrative style. The notion was to start with ambitious standards and strict deadlines, stimulate crash programs of engineering work in Detroit, and then relax the timetable as appropriate when deadlines drew near. They refused to accept arguments that passive protection was infeasible or ill-advised.

One of the remarkable aspects of this period was the positive chemistry in the relationship between Volpe and Toms. Detroit could not drive a wedge between them. By the spring of 1971 their leadership had achieved some important accomplishments, such as vastly improved air bag designs and development of potentially promising alternatives to the air bag. They had also sustained NHTSA's reputation for independence and toughness in its role as a "technology-forcing" safety agency.

On the negative side, the Volpe-Toms style instigated substantial opposition and mistrust within the auto industry. Their deadlines were perceived as naive and unrealistic, and their pro–air bag posture was perceived as extremist. Throughout this period there were fears that this manufacturer or that supplier was getting ahead in the competitive scramble by exploiting personal contacts in the government. A somewhat more dispassionate and moderate position—for example, a phase-in of mandatory passive restraints along the lines suggested by Ed Cole—might have engendered more trust and credibility for NHTSA within the industry and done more in the long run to enhance safety.

Endnotes

1. R.M. Kemmerer, R. Chute, D.P. Haas, and W.K. Slack, "Automatic Inflatable Occupant Restraint System," SAE Paper Number 680033, Automotive Engineering Congress, Detroit, Michigan, January 8–12, 1968.
2. "Air Bag Concept for Safety Looks Good," *Steel* (December 30, 1968), pp. 9–10.
3. "Air Cusion Device Called Key to Safety in Mini-Car Crashes," *New York Times* (April 20, 1969), p. 92.
4. "Air Bag Concept for Safety Looks Good," p. 10.
5. Interview of David Martin, September 3, 1987.
6. "Air Bag Concept for Safety Looks Good," p. 10.
7. "Auto-Ceptor," *Steel*, December 30, 1968, p. 9.
8. Ibid.
9. "Air Cushion Device Called Key to Safety in Mini-Car Crashes," *New York Times* (April 20, 1969), p. 92.
10. Interview of David Martin, September 3, 1987.
11. "Air Cushion Device Called Key to Safety in Mini-Car Crashes," p. 92.
12. "Air Bag for Safety in Car Crashes Packs a Punch," *New York Times* (June 17, 1969), p. 49 (quoting Ford engineer Robert Marshall).
13. Interview of Lowell Dodge, August 13, 1987.
14. Interview of Joan Claybrook, July 24, 1987.
15. Interview of Robert Carter, August 17, 1987.
16. "Air Bag for Safety in Car Crashes Packs a Punch," p. 49; *Federal Register* 34:11,148 (1969); "U.S. Plans to Order 'Air Bags' in Autos," *New York Times* (July 3, 1969), p. 12.
17. Interview of David Martin, September 3, 1987; "New Safety Device Pres-

sures Detroit," *Business Week* (July 12, 1969), pp. 45-46; "Auto Safety: Passive Systems Pushed," *Steel* (October 20, 1969), pp. 7–8.

18. "New Safety Device Pressures Detroit," p. 46.
19. "Ford to Introduce a Safety Air Bag on the '71 Mercury," *New York Times* (August 22, 1969), p. 70.
20. Ibid.
21. "A Delay Is Sought on Air Safety Bag," *New York Times* (August 29, 1969), p. 58.
22. "Chrysler Is Wary of Air Bag Device," *New York Times* (August 27, 1969), p. 86; "Expert Sees Air Bag Delay," *Automotive Industries* (November 15, 1969), p. 98.
23. Ibid.
24. "A Delay Is Sought on Air Safety Bag," p. 58.
25. "Happiness Could Be an Air Bag," *New York Times* (August 31, 1969), p. III-2.
26. Ibid.
27. "New Safety Device Pressures Detroit," *Business Week* (July 12, 1969), pp. 45–46.
28. "Bag that Protects Motorists," *New York Times* (June 28, 1969), p. 43.
29. "Auto Safety Program of the Administration Apparently Has a Chief," *Wall Street Journal* (September 12, 1969), p. 15; interview of Robert Carter, September 1987; "Auto Safety Nominee," *New York Times* (December 6, 1969), p. 20.
30. Interview of Robert Carter, August 17, 1987.
31. Interview of Douglas Toms, July 20, 1987.
32. Ibid.
33. Ibid.
34. "Auto Safety Nominee," p. 20; "Volpe's Consultant Offers Plan to Upgrade Auto Safety Office," *New York Times* (November 27, 1969), p. 74.
35. Interview of Joan Claybrook, July 24, 1987.
36. Interview of Ralph Hitchcock, September 23, 1987.
37. *Federal Register*, 35:7,187 (1970).
38. "Air Bags: DOT Readies Decree for Big Market, Huge Headache," *Industry Week* (April 6, 1970), pp. 21–22.
39. Ibid.
40. Interview of Douglas Toms, July 20, 1987.
41. Interview of Robert Carter, August 17, 1987.
42. "Air Bag Parlay Called Remarkable," *Automotive Industries* (June 15, 1970), pp. 37–39; Joseph M. Callahan, "Industry May Sink Air Bag Project," *Automotive Industries* (June 15, 1970), p. 17.
43. Susan J. Tolchin, "Air Bags and Regulatory Delay," *Issues in Science and Technology* 1 (1984), p. 70.
44. Joseph M. Callahan, "First Air Bags Predicted in 1974," *Automotive Industries* (June 1, 1970), p. 17.
45. "Air Bag Parlay Called Remarkable," pp. 37–39.
46. Interview of Douglas Toms, July 20, 1987.
47. Interview of Joan Claybrook, July 24, 1987.
48. "Carbuilders Balking at Air Bags," *Industry Week* (June 1, 1970), pp. 71–72.

49. "Air Bag Parlay Called Remarkable," pp. 37–39.
50. Interview of David Martin, September 3, 1987.
51. "Air Bags Got Bugs, Say Automen," *Iron Age* (May 21, 1970), p. 23; "Air Bag Plan Seen Delaying Car Safety Gains," *New York Times* (September 7, 1969), p. 84.
52. Interview of Robert Carter, August 17, 1987.
53. "Air Bag Parlay Found Remarkable," pp. 37–39.
54. Joseph M. Callahan, "Industry May Sink Air Bag Project," *Automotive Industries* (June 15, 1970), p. 17.
55. 'Air Bag Parlay Called Remarkable." pp. 37–39.
56. Ibid.
57. "Volpe's Safety Agenda," *Automotive Industries* (July 15, 1970), p. 23.
58. Ibid.
59. "GM Proposes Timetable for Air Safety Cushion," *Machine Design* (July 23, 1970), p. 10.
60. Interview of Douglas Toms, July 20, 1987.
61. Interview of Robert Carter, August 17, 1987.
62. Ibid.
63. Letter from Congressman Adam Benjamin to Secretary John Volpe, June 23, 1970; also letter from Senator Frank Moss (D–Mass.) to Secretary John Volpe, June 22, 1970.
64. Interview of Douglas Toms, July 20, 1987.
65. "Blanket Battles Bag in Safety Bout," *Iron Age* (July 16, 1970), p. 21.
66. Interview of Robert Carter, August 17, 1987.
67. "A Faster Road to A Safer Car," *Business Week* (October 31, 1970), p. 23.
68. *Federal Register* 36:19,266 (1971).
69. *Federal Register,* 35:16,928 (1970).
70. "CAL to Evaluate Air Bags," *Automotive Industries* (November 15, 1970), p. 92.
71. "Head Bags Next," *Industry Week* (April 6, 1970), p. 22.
72. Interview of Robert Carter, August 17, 1987.
73. Interview of James Hofferberth, August 17, 1987; interview of Ralph Hitchcock, July 23, 1987.
74. "Improved Air Bags May Remove Obvious Hazards," *Product Engineering* (February 1, 1971), p. 13.
75. Ibid.
76. "Is There Alternative to Air Bag?" *Industry Week* (February 8, 1971), p. 63.
77. "Air Bag Alternatives: Blankets and Nets and Court Action?" *Automotive Industries* (March 1, 1971), p. 22.
78. Ibid.
79. "Eaton Bucks Air Bag Resistance," *Industry Week* (March 20, 1971).
80. Interview with Douglas Toms, July 20, 1987.
81. "Air Bag Plan Seen Delaying Car Safety Gains," *New York Times* (September 7, 1969), p. 84; "Will Chrysler Deflate the Air Bag?" *Iron Age* (February 11, 1971), p. 23.
82. Interview with Douglas Toms, July 20, 1987.
83. "Air Bags: On the Way to Reality," *Iron Age* (March 25, 1971), p. 21.

84. "Eaton Bucks Air Bag Resistance," *Industry Week* (March 20, 1971).

85. Interview of Douglas Toms, July 20, 1987.

86. "Seat Belt Man Talks Air Bags," *Industry Week* (May 3, 1971), pp. 75–76.

87. Interview of Douglas Toms, July 20, 1987; interview of Robert Carter, August 17, 1987.

88. Interview of Ralph Hitchcock, July 23, 1987.

89. "Air Bags: On the Way to Reality," p. 21.

90. Ibid.

91. Interview of Lowell Dodge, August 13, 1987.

92. Letter of Ralph Nader to Secretary John A. Volpe, March 8, 1971.

93. "The Air Bag Faces a Showdown Fight," *Business Week* (August 14, 1971), pp. 74–75.

Chapter 4

THE NIXON WHITE HOUSE AND FORD'S "BETTER IDEA"

The basic axiom of pluralism is that there is no single source of sovereign power. The lofty ambitions of actors at one power center, noble as they may be, can be blocked by opponents at other power centers through an overt assault, persuasion on the merits, or subtle inducement. A leader of one source of power must therefore take into account the perspectives, incentives, and proclivities of political opponents with access to other sources of power.

A case can be made that John Volpe and Douglas Toms behaved as if their "technology-forcing" authority was some form of unchecked authoritarianism. They pressured the auto industry so hard on the air bag from 1969 to 1971 that they virtually invited laggard manufacturers to appeal the wisdom of their technology-forcing rule to other power centers. When Ford and Chrysler took NHTSA's position to other authorities, it became apparent that Volpe and Toms did not hold all the cards. Even though Congress played a supportive (and perhaps instigating) role, NHTSA was trumped in both the White House and the federal courts. Given how vulnerable NHTSA was to White House and judicial disapproval, one can argue that a more modest "technology-inducing" posture would have been more credible in the eyes of industry and therefore would have done more for safety.

The danger in provoking backlash from other power centers is that the technological or policy options imposed by these authorities may prove to be less desirable than taking more modest steps in the first place. NHTSA possesses much more technical competence on safety matters than does Congress, the White House, or the federal courts. By allowing themselves to be bypassed by Ford

Motor Company, Volpe and Toms inadvertently contributed to ill-advised decisions, creating what has proven to be the most embarrassing and politically damaging experience in the agency's history. As we shall see, the experience was extremely harmful to America's pursuit of safety.

A Corporate Flip-Flop

One of the critical turning points in the air bag controversy was the flip-flop that occurred at Ford Motor Company in 1970. During the late 1960s Ford was the industry's leading proponent of the air bag. Ford's vice president of engineering, Herbert Misch, was cautiously optimistic about the technology.[1] Matters changed quickly. By June 1970 Ford was sponsoring national newspaper and magazine advertisements that criticized the air bag and promoted the ignition-interlock device as a superior alternative.[2] The explanation for this turnaround lies in the growing skepticism of Chairman Henry Ford II and President Lee Iacocca.

The air bag idea evolved in different ways at Ford and at GM. At General Motors the air bag program was primarily a "top-down" initiative, arising from the interests of President Ed Cole. At Ford the air bag was more of a "bottom-up" initiative which evolved under the stewardship of engineer Jack Pflug, who was project officer of Ford's exploratory program with Eaton, Yale and Town, Inc.

When the air bag was a small research and development effort at Ford, the attitude of top management was perhaps indifferent and in any case not critical. When the air bag began to look like a billion-dollar regulatory compliance program, the attitudes of Chairman Henry Ford II and Lee Iacocca were decisive. The thinking of Iacocca was especially critical because he had some experience in safety and his power and influence at Ford were on the rise.

Enter Lee Iacocca

As a self-described "safety nut," Lee Iacocca was not your stereotypical corporate executive with callous attitudes toward consumer safety. As early as 1955 he was part of a marketing group at Ford that promoted the industry's first large-scale safety package. Working under the leadership of Ford executive Robert McNamara, this group offered a safety package on 1956 models comprised of lap belts, safety door latches, sun visors, crash padding, and a deep-dish steering wheel.[3]

This multimillion-dollar safety campaign proved to be a failure, at least in the eyes of most industry officials. For example, only 2 percent of buyers took the seat belt option. Ford abandoned the safety campaign when arch-rival Chevrolet stole its market by advertising high-powered V-8 engines and jazzy wheels. Chevrolet widened its sales lead over Ford from 70,000 cars in 1955 to nearly 200,000 in 1956.[4] Nader has argued that some of Ford's safety options were actually marketing successes and were terminated prematurely.[5] It is therefore possible that Ford's sales would have been even worse if the safety campaign had not been launched.

Ford's embarrassing experience apparently conditioned Iacocca—and much of the industry—to believe that "safety doesn't sell." In Iacocca's words, "safety is a pretty poor marketing device, which is why the government has to get involved."[6] What this ultimately meant was that Iacocca was inclined to look for inexpensive approaches to meeting safety goals because most buyers were seen as unwilling to pay a premium for extra safety. In light of his experience, it is no surprise that Iacocca was an opponent of the air bag. Like many auto industry officials, Iacocca argued that the seat belt was a cheaper, simpler, and more reliable device than the air bag. The problem was that most motorists weren't buckling up. To correct the problem Iacocca became an early supporter of mandatory seat belt wearing laws. Yet he found few political allies to lobby for this approach, and he did little himself to promote such laws.[7]

When Iacocca was appointed president of Ford Motor Company in December 1970 by chairman Henry Ford II, it signaled the eventual demise of Ford's faltering air bag program. Iacocca had little reason to believe that Ford could do a better job on air bags than Ed Cole at GM. He therefore searched for a different corporate response to the political demand for auto safety.

The "Better Idea"

In response to Secretary Volpe's upcoming passive-restraint regulation, Iacocca worked with a group of Ford engineers on a different idea: the starter-interlock system (called the "interlock" for short).[8] The interlock was designed to prevent the starter from working unless the driver and passenger were buckled.

Ford was not the only company working on the interlock. In the summer of 1970 two small manufacturers (Peugot and Renault) requested that NHTSA regard interlocks as an acceptable substitute for passive restraints. Several seat belt manufacturers (Takata Kojyo and Irvin Industries) also argued to NHTSA that interlocks should be permitted—at least until air bags were fully developed and proven.[9]

The expectation of interlock advocates was that the system would cause "a quantum jump in safety belt use."[10] The interlock was also less expensive than the air bag and could be installed relatively soon without years of development work.

After several meetings in Washington, Iacocca learned that federal officials, including Administrator Toms and Secretary Volpe, disliked the interlock because it was not a "passive" system. The first administrator of the safety bureau, Dr. William Haddon, had exerted a profound influence on the thinking of agency officials with his notion of safety devices that would work without requiring any action on the part of vehicle occupants.

In March 1971 Volpe and Toms issued a final rule that required front-seat passive protection in new cars by August 15, 1973. Passive protection for rear-seating positions was delayed until August 15, 1975, because air bag development for the rear seat was not progressing adequately. Toms issued an official "interpretation" of "passive" restraints, in May 1971, that explicitly excluded "forced action" systems such as the interlock.[11] It appeared that the interlock issue had been settled.

Appeals to the White House

In late 1970, several auto executives (Chairman Richard Gerstenberg of GM, Chairman Lynn Townsend of Chrysler, and Chairman Henry Ford II and President Lee Iacocca of Ford) went directly to President Richard Nixon to voice their concern about the ambitiousness of NHTSA's passive-restraint rule and EPA's tailpipe emission rules. Toms explains his thinking at the time:[12]

> *We knew that the Detroit people were after our butts. We also knew that the air bag development programs were not progressing as rapidly as we had hoped. Although we were annoyed that they were trying to run around us, we sensed that a more moderate policy position was going to be necessary.*

What Toms did not realize was that Iacocca and Henry Ford II were also maneuvering independently of the rest of the industry to get the White House to undo NHTSA's passive-restraint rule. In a meeting at the White House with President Nixon and John Ehrlichman on April 27, 1971, Henry Ford II and Lee Iacocca lobbied for help in repealing the passive-restraint rule. This meeting was apparently unknown to Volpe and Toms at the time and was not made public until a decade later (December 1982), when one of Nixon's White House tapes was disclosed in a wrongful death trial involving the Ford Pinto.[13] The tape indicates that the result of the meeting was that John Ehrlichman, Nixon's domestic affairs adviser, became Ford's White House contact on the passive-restraint issue. Ehrlichman promised to talk with Volpe.

Ehrlichman and White House aide Peter Flanigan exerted pressure on Volpe to "back off" on passive restraints. Ehrlichman met personally with Volpe and sought repeal of the passive-restraint rule, but Volpe resisted.[14] A technical briefing on the air bag's effectiveness by Volpe's associates did nothing apparently to sway Ehrlichman.[15]

Ralph Nader claims that Volpe was in effect ordered by the White House to delay the passive-restraint ruling. Nader also speculates that Volpe relented because his job was at stake.[16] Volpe has denied this speculation in an interview for the CBS program "60 Minutes." Volpe acknowledges that there were communications with the White House officials during this period but insists that he was only asked to reconsider his March 1971 decision.[17]

Toms remembers the events somewhat differently than Nader. "We knew our deadlines for passive restraints were looking unrealistic and hence the White House requests only confirmed our inclination to take a more moderate position," says Toms.[18] Toms "cautioned" Volpe against the interlock on grounds of adverse public reaction but "both of us saw that it was hard to deny Detroit a shot at safety during the interim when the air bag was still not ready."[19] Because Volpe and Toms were "realists," they reluctantly went along with the interlock idea.[20]

Regardless of the precise reasons, Iacocca got what he wanted. On October 1, 1971, NHTSA proposed a modified rule that required front seats of new cars to be equipped with either interlocks or passive restraints by August 15, 1973. The requirement for passive protection was delayed two years (until August 15, 1975) to allow more time for air bag development.[21]

NHTSA's stated rationale for delaying the rule was that the earlier deadline might have caused "extreme dislocations" and "financial hardships" in the industry.[22] Administrator Toms explained to reporters that "we've been holding the industry's feet to the fire on this but we are aware of their problems. I don't think any responsible engineer is opposed to the air cushions, it is really an issue of time."[23]

The modified rule permitting interlocks, which was adopted in final form in February 1972, represented quite a victory for Ford.[24]

Ford's heavy-handed strategy in dealing with NHTSA stood in striking contrast to the hands-off policy at GM. Toms explains:[25]

> *Ed Cole at GM believed that federal regulation was necessary for safety. He was inclined to roll with the regulatory process as long as it was applied equally to all firms. Since GM was the largest firm and had the best technical resources, Cole knew that technology-forcing rules would ultimately help GM stay on top.*

Internal agreement about firm policy was also greater at Ford than at GM. Henry Ford II and Lee Iacocca were united against the air bag, whereas GM Chairman Richard Gerstenberg and Ed Cole did not see eye to eye on this one. Gerstenberg was perceived as "skeptical" of the air bag, though he let Cole manage the issue.[26]

First Round of Litigation

The two-year delay in the passive-restraint requirements did not win favor among any of the interested parties. Ralph Nader sued NHTSA in the U.S. District Court of the District of Columbia on the grounds that the delay was instigated by White House political pressure. This suit failed when the court, after reviewing White House materials, refused Nader access to White House memoranda. Chrysler, American Motors, Ford, Jeep Corp., Volkswagen, Audi, and the Automobile Importers of America had sued the agency six months earlier in the U.S. Court of Appeals for the Sixth Circuit in Cincinnati, Ohio. They made a variety of procedural, statutory, and technological arguments against the agency's rule on passive restraints.

The Sixth Circuit was believed to be a sympathetic forum for the auto industry because it covered much of the industrial Midwest. The case was critical because of the immediate stakes and the potential precedents it might set for NHTSA's future. To present its case, the Ford Motor Company hired a blue-chip Washington law firm, Wilmer, Cutler and Pickering. Lloyd Cutler had been a key player for industry in the final negotiations of the 1966 Safety Acts.

The Center for Auto Safety, led by Lowell Dodge, was particularly concerned that the agency win the "technology-forcing" issue in this litigation. Dodge submitted an amicus brief supporting the agency's authority to set requirements that go beyond feasible technologies. The industry, in contrast, was arguing that NHTSA was restricted to regulating on the basis of already-proven technology.

The only major manufacturer that did not sue NHTSA was General Motors, which was also the lone carmaker still committed to the air bag. According to Edwin H. Klove, Jr., top engineer at GM's Fisher Body Division, the air bag was "the most promising passive-restraint system known to GM."[27] While tolerant of the company's air bag program, GM Chairman Richard Gerstenberg was opposed to NHTSA's modified rule as a matter of public policy because it did not allow a long enough phase-in period to conduct large-scale field tests.[28] Yet GM did not join Ford and Chrysler in their litigation. GM had much to gain *competitively* from an ambitious rule.

Haddon's New Role

After leaving the job of NHSB director, William Haddon persuaded the nation's major property and casualty insurance companies to launch an expanded research and communications organization. The Insurance Institute for Highway Safety (IIHS) in Washington, D.C., was considered a rather sleepy place until Haddon became its president. It was basically "a conduit for insurers to transfer funds to traditional state and local safety groups."[29] Its philosophy was: "We don't tell Detroit how to build cars; they don't tell us how to write insurance."[30] Haddon brought with him from NHSB a talented public relations expert, Benjamin Kelley. For the next decade Haddon and Kelley worked diligently to tarnish Detroit's public reputation on safety issues and to keep the air bag issue on the national agenda.[31] IIHS became the principal source of facts, numbers, and excellent film footage about air bags.

Haddon was not loved in Detroit, but he shared GM's concern that the air bag needed a large-scale field test. Haddon felt that the government was risking a premature rejection of the air bag technology by in effect requiring that manufacturers move directly from proving-ground tests to the marketplace.[32] An overly ambitious rule could, Haddon warned, "blow out of the water for a long time, maybe even permanently, an extremely promising system."[33]

NHSB officials had revealed plans in May 1970 to undertake a five-year field test program of 12,500 air bag equipped vehicles. This project never got off the ground, in part because the government was convinced that manufacturers would conduct their own field tests. But the field-testing plans of Ford were slashed by

Iacocca to less than 1,000 vehicles, and GM's ambitious plans for field testing were delayed by Cole due to engineering problems. When NHTSA issued its modified order in February 1972, the air bag still had not been exposed to any significant real-world testing.[34] Thoughtful air bag advocates, such as Haddon, knew that such an innovative technology needed to be introduced into the fleet gradually .

Liability Fears

As the tide seemed to turn against the air bag, Toms and his associates discussed strategies to reduce the reluctance of manufacturers to offer air bags. Toms knew from his visits to Detroit that "the potential liability exposure from air bags scared the hell out of industry lawyers and bean counters."[35] Air bags could increase litigation of all sorts: accusations of failure to deploy, of inadvertent deployment, of deployment-induced injury, and so forth. Even false claims would be expensive to defend against. This fear of liability, Toms felt, was generating "the most stubborn resistance to air bags in Detroit."[36]

In the spring of 1972 Toms arranged a one-on-one meeting with Senator Warren Magnuson (D–Wash.) to discuss the liability issue. A legislative solution was required because NHTSA had no leverage over common-law rulings by judges. Toms and his associates had discussed options such as caps on jury awards and creation of a legal defense for manufacturers based on compliance with federal safety standards.

Senator Magnuson informed Toms that liability relief for Detroit would go nowhere in Congress. The legislative mood was not suitable and the issue had little political appeal. Toms recalls thinking at the time, "There are too many lawyers in the Congress."[37] Toms searched in vain for a strategy to lessen corporate fears of lawsuits involving the air bag technology.

Adverse Media Publicity

As the Sixth Circuit Court of Appeals deliberated about the conflicting legal claims against Toms and Volpe, both the NHTSA and the air bag became targets of negative press coverage. A report by the White House Office of Science and Technology (OST), "Regulatory Effects on the Cost of Automotive Transportation" (RECAT), was released several days prior to oral argument at the Sixth Circuit.[38] *Business Week* reported that the RECAT study "clobbered EPA and NHTSA" in an attempt to "weaken the regulatory powers that permit these two agencies to set standards."[39] The report was actually more critical of emission rules than safety rules, but it did call into question the cost-effectiveness of air bags. Mandatory seat belt wearing laws were cited in the study as a potentially promising alternative.

Lowell Dodge of the Center for Auto Safety uncovered an internal White House memorandum that had directed OST to undertake the RECAT study.[40] As a result, Ralph Nader charged publicly that the RECAT study was motivated by President Nixon's desire to "win friends" in Detroit during an election year.[41] The chairman of the RECAT panel, Dr. Lawrence A. Goldmuntz, denied this charge in subsequent congressional testimony, stating that the report was initiated by himself despite White House reservations.[42]

The adverse publicity from the RECAT report was compounded by two public incidents involving the air bag. In the first public test of the "Experimental Safety Vehicle" (a NHTSA-supported project run by Fairchild Industries in Phoenix, Arizona), an air bag failed to inflate when the vehicle hit a fixed barrier at 50 mph. The problem was apparently faulty soldering of the electrical leads running to the air bag control box. When the lines were later connected properly, the bags inflated. Lawyers for Ford used news clips from the incident to bolster their case against the air bag before the Sixth Circuit. Proponents of the air bag insisted, without much press attention, that this type of defect might have been prevented by normal quality control systems.[43]

Later in the summer of 1972 a team of scientists at Wayne State University supervised a public test of the air bag for the National Motor Vehicle Safety Advisory Council, an independent adviser to the U.S. Department of Transportation. The Eaton air bag failed to deploy in the test, an event that received much notoriety in the press. According to Eaton engineers, the Wayne State team used an obsolete and defective gas generator that was not tested prior to use.[44] Lowell Dodge commented: "There is no evidence that this failure was intentionally rigged. The degree of negligence on the part of those conducting the experiment, however, was so great as to maximize the chances of failure."[45] Career NHTSA officials insist, however, that the failure was inadvertent. The suspicions of sabotage are a clear indication of the mistrust that characterized this period.

The adverse publicity about air bags created quite a predicament for technical people at NHTSA. While opponents of air bags—especially suppliers of belts—did not hesitate to go public with criticisms of air bags, engineers at NHTSA were reluctant to go public about the limitations of belts. Engineer James Hofferberth of NHTSA explains:[46]

> *We in the agency were very careful not to publicize the possible problems and inherent limitations of belts, because of the possibility that people would view the inherent limitations of belt systems as yet another ill-founded excuse not to wear them.*

As a result, the public heard a very mixed message about air bags along with a highly favorable report about belt systems.

Buckle-Up Laws Considered

While the air bag became embroiled in litigation and public controversy, the case for mandatory seat belt wearing laws was beginning to be made. Victoria, a province of Australia, was the first jurisdiction in the world to adopt such a law. Belt-wearing rates rose from 20 to 80 percent, and the province reported a 13 percent decline in occupant fatalities in rural areas and a 24 percent decline in urban areas within the first nine months after the law was enforced.[47]

The report of Victoria's experience was followed by a National Safety Council endorsement of such laws for the United States. Although the American Automobile Association opposed such laws on grounds of infringement of free choice, both Lee Iacocca of Ford and GM Chairman Richard Gerstenberg made public speeches in favor of belt use laws. That summer Volpe approved a plan to include mandatory seat belt wearing laws as part of NHTSA's periodic evaluation of state highway safety plans under the 1966 Highway Safety Act. Although twelve states considered belt use laws in 1972, none were enacted into law.[48]

Some air bag advocates feared mandatory belt use laws, believing that such enactments might undercut the rationale for passive-restraint regulation. But, according to NHTSA's James Hofferberth, that is not the primary reason the agency "didn't actively pursue mandatory use laws." Agency officials were convinced, Hofferberth explains, "that the chances of getting laws passed in the states were low," and if the laws were passed and enforced vigorously, "the ensuing public backlash [against government] would do more harm than good to the long-term safety initiative."[49]

Momentum Behind Automatic Belts

Many supporters of non–air bag technologies felt that NHTSA's rules were designed to discourage alternatives to air bags. As early as July 1971 the agency proposed a new rule requiring that automatic belts be nondetachable (to encourage use) with "emergency-release" mechanisms to allow escape from the vehicle after a crash.[50] The design restrictions on automatic belts were viewed as evidence of NHTSA's bias toward air bags. Toms in particular was known as "an unrelenting air bag booster."[51] Many industry officials saw the passive-restraint rule as a de facto air bag requirement.

At the NATO Conference on Passive Restraints in June 1972, the most significant revelation seemed to be NHTSA's evolving position on "passive belt systems." Administrator Toms acknowledged that some belt-based concepts might be developed to satisfy the passive-restraint rule. In particular, the agency highlighted Volkswagen's work on automatic shoulder belts and knee pads.[52] To the surprise of many conference participants, Toms made clear that he was serious about passive belts, citing remarkable improvements in the performance of such belts in 30 mph crash tests. For small cars, he saw passive belts as an especially attractive option.

The agency's permissive position on passive belts arose in spite of technical disagreements among career officials at NHTSA. Robert Carter (motor vehicle programs) was very concerned about the crash performance characteristics of automatic belts, referring to "killer belts" that in crash tests would cut dummies in two. In Carter's opinion, belts should be subjected to even tougher performance standards than air bags. Larry Schneider (chief counsel) and James Hofferberth (senior engineer) acknowledged the technical concerns but argued that the rule would be more palatable if more compliance options were made available to the industry. Toms sided with Schneider and Hofferberth and set a critical precedent in favor of a permissive attitude toward automatic belts.[53]

Toms also said at the NATO conference that he anticipated petitions from manufacturers requesting an extension of the permissibility of interlocks through the 1976 model year. The agency's response to such petitions would depend, he said, on the usage rates achieved by the interlock. As a target, he set 85 percent use as a level that might justify delay of the passive-restraint requirements.[54]

Whether air bags should substitute for belts in the long run was a matter of heated dispute. Engineers at GM and NHTSA saw the possibility of making lap belts an optional addition to air bags; Ford and Chrysler engineers insisted that it was "dishonest" to maintain that air bags could completely substitute for seat belts.[55] In the face of these technical disputes, each manufacturer was struggling to formulate a coherent corporate strategy.

The "Rubber Yardstick"

The litigation of Volpe's February 1972 rule had a stifling effect on air bag development. One trade journal commented that manufacturers "put [the air bag] on the back burner until the court rules."[56] The ultimate 2–1 decision by the Sixth Circuit panel offered no clear victory for either side.[57]

The judges upheld DOT's authority to mandate passive restraints and approved DOT's technological and economic analyses of the air bag. Ralph Nader and Lowell Dodge were encouraged because the panel acknowledged NHTSA's "technology-forcing" authority.[58] Yet the panel suspended the passive-restraint requirements indefinitely on the grounds that NHTSA could not measure manufacturer compliance with the standard in "objective" terms. Two of the three judges (John Peck and Paul Weick with William Miller dissenting) were convinced that the dummies used in crash tests did not provide predictable and consistent results.[59] Dodge recalls that "none of the attorneys in the case dreamed that the dummies would decide the case."[60] The court ordered the agency to improve compliance tests before issuing another rule.

One of Ford's veteran legal experts recalls that NHTSA's test dummy was "the least appealing issue in the case from the standpoint of legal scholarship." To emphasize its importance to Ford's outside attorneys, he referred to NHTSA's "rubber yardstick," reflecting the belief of Ford's engineers that compliance measurements based on dummies were totally unpredictable.[61]

The agency's loss on the testing issue was not a complete surprise to career scientists at NHTSA. According to a NHTSA engineer in the rule-making office, "the problem in the agency was that turnover among our technical staff caused this aspect of the standard to fall through the cracks."[62] The court was apparently correct: The dummies used in crash testing were not yet well enough calibrated to provide consistent head and neck measurements. A NHTSA engineer acknowledges that "it was a fatal technical error on our part because by neglecting this solvable problem we opened the door for Chrysler's lawyers."[63]

The victory was especially rewarding for Chrysler. As the industry's weakest company, Chrysler stood to lose the most from new regulatory requirements because it was far behind General Motors and Ford in air bag development. Moreover, Roy Haeusler was convinced that belt use was the fundamental challenge while air bags were at best a promising research idea. Haeusler had worked with Chrysler's attorney Robert Sorenson to expose the technical weaknesses in NHTSA's position.[64]

Meanwhile President Nixon began his second term in the White House by asking for the resignations of all his cabinet officials. Apparently determined to gain greater control over "the bureaucracy" in his second term, Nixon directed White House aides Robert Haldeman and John Ehrlichman to purge the Cabinet of anyone who was not a Nixon loyalist. Among others, Volpe at DOT and George Romney at HEW became casualties of Nixon's order, although Volpe was subsequently appointed ambassador to Italy.

The new secretary of transportation was Claude Brinegar, a senior vice president of Union Oil Company of California. Brinegar was a quiet and studious economist with no apparent interest in auto safety. He confronted considerable difficulty in finding a replacement for Toms, who had resigned with Volpe. The loss of Toms was a big blow to NHTSA because he had demonstrated an ability to mobilize and utilize NHTSA as an organization.

GM's Wavering Strategy

A critical player at GM during this period (1972–1974) was the director of safety engineering, Louis Lundstrom. NHTSA officials describe him as a sound engineer and a man of integrity. His attitude toward the air bag was neutral: He wanted to wait and see what the field experience would show. Lundstrom charged George Smith, an air bag critic, with the job of monitoring and evaluating GM's fleet of 1973 Chevy's with air bags. This fleet made some GM engineers very nervous because the cars were not equipped with belts (though they were later retrofitted with them).[65] Ford officials were amazed that GM would let these cars on the highways, for they believed that air bags alone could not possibly match the occupant protection offered by lap/shoulder belts.[66]

Throughout 1972 GM was gearing up to launch a large-scale, real-world air bag testing program. This effort was of critical interest to engineers at NHTSA because GM was the lone manufacturer still committed to early introduction of the air bag. Although Ford equipped 831 Mercurys in 1971–72 with Eaton's air bag system, by the end of 1972 the management at Ford had postponed its air bag offerings in favor of the interlock until research engineers could resolve lingering technical problems. GM began by launching a field test of 1,000 air bag–equipped 1973 Chevys, but this fleet was too small to provide statistically reliable data about air bag performance.[67]

Allstate Insurance Company played a supportive role in this period by purchasing small fleets of the 1972 Mercurys and the 1973 Chevys with air bags. Working with Hal Needham, a noted Hollywood stunt man, Allstate was the first to perform live human crash tests with air-bag-equipped vehicles. The films of these tests, which captured Needham plowing into a brick wall at 30 mph and then walking away, were later used on Capitol Hill to build support for the air bag among senators and congressmen. Allstate also made print and TV ads using the live crash films, which were ultimately carried by ABC and NBC (but not CBS).

In January 1973 GM placed 100,000 orders for Eaton's air bag sensor system, signaling the company's commitment to selling cars with the new safety device. The plan was to offer air bags as an option on 1974 models of the Buick Electra and Riviera, the Oldsmobile 98 and Toronado, and most Cadillacs.[68]

In April 1973 NHTSA proposed a revision in its test dummy specifications in response to the shortcomings cited in the Sixth Circuit's decision.[69] Engineers at the agency had been working hard to solve this problem, using knowledge generated by General Motors. The dummy design proposed by the agency, called the "Hybrid II Dummy," was developed by General Motors Corporation, Alderson Research Laboratories, and Sierra Engineering Co.[70]

In the months following NHTSA's proposal, a heated public debate developed about whether the agency should reestablish its requirement that all 1976 cars be equipped with passive restraints. The Center for Auto Safety petitioned the agency to reestablish the deadline because problems with the test dummy had been resolved, and several insurance industry spokespersons backed CAS's position. Although many career officials at NHTSA supported this idea, the agency had no permanent head to rule on such a key issue.

In July 1973 General Motors President Ed Cole wrote to Secretary Brinegar requesting that the deadline for passive restraints be delayed "indefinitely—or at least until an appropriate phase-in can be established." Cole noted that the company could not implement its earlier plan to install air bags in 100,000 1974 models. The delays in GM's air bag program were attributed in part to uncertainty about NHTSA's test specifications, which (though proposed) were still not published in final form.[71]

Cole's letter was a sign that GM's commitment to the air bag was softening. There were other signs as well. A new GM study by engineers Richard Wilson and Carl Savage questioned the relative effectiveness of air bags. This study of a sample of 706 real-world crash fatalities concluded that air bags alone would have provided less protection to occupants than either lap/shoulder belts or the air bag/lap belt combination (assuming belts are worn).[72] This study undercut Cole's desire to completely replace belts with air cushions. A policy speech delivered by David E. Martin, an automotive safety engineer at GM, emphasized that consumers should retain some freedom to choose restraint systems. He criticized "safety authorities" who would "take this choice out of their [consumer's] hands through [mandated] passive restraints."[73] Martin recalls telling GM President Ed Cole, about this time, that "the only thing that would make an air bag enthusiast out of me is an automatic belt."[74]

The decision of GM to abandon the air bag–only system in favor of an air bag/lap belt combination did not come about easily. GM's safety engineers enjoyed the support of Executive Vice President Pete Estes in their effort to persuade Cole to add a lap belt. David Martin describes Cole as a "man of supreme confidence" who only very grudgingly accepted the conclusion that air cushions must be accompanied by a lap belt.[75]

Engineers at NHTSA were sensitive to Cole's criticism of their delays on the testing issue. They quickly published final test dummy specifications in 1973 and permitted manufacturers to install passive restraints. The final test specifications were revised slightly in order to satisfy concerns of GM engineers about whether Hybrid II would conform to NHTSA's specifications.[76] The announcement accompanying the final rule made clear that NHTSA was eager for GM to proceed with its plans to offer air bag–equipped cars. This move seemed effective at the time in view of GM's subsequent announcement to produce 50,000 air bag–equipped cars beginning January 1, 1974.[77] No decision was made about whether the 1976 model deadline for mandatory passive protection should be delayed.

Reaffirming Interlocks

During the year following NHTSA's February 1972 decision to permit interlock devices (instead of passive restraints) on 1974 models, pressure mounted for reconsideration. In December 1972 GM officials told NHTSA that, "we believe that use of the starter interlock should be avoided because of the high risk of customer reaction against the use of belt restraints."[78] Chrysler and Fiat joined GM in formally petitioning NHTSA to withdraw the interlock mandate.

The National Motor Vehicle Safety Advisory Council (NMVSAC) also urged Secretary Brinegar in March 1973 to drop the interlock rule. These recommendations were predicated primarily on fears that consumers would dislike the interlock, but concerns were also raised about the reliability and safety of the device. GM and Chrysler urged the agency to allow manufacturers to use less intrusive buzzer-light reminder systems.[79]

The agency denied these petitions on the grounds that they were "necessarily speculative." On the subject of reliability, the agency insisted that "the degree of reliability of any system is a function of the manufacturer's own design and quality control."[80] Secretary Brinegar wrote NMVSAC and explained that DOT had decided "after a thorough review of the issues that the public benefits from the ignition interlock system will outweigh the costs."[81]

Enter James Gregory

Although Brinegar remained decisive in favor of the interlock, he made no decision about when the passive-restraint requirements would take effect. Meanwhile, Brinegar was having difficulty finding a suitable replacement for Toms.[82] NHTSA lacked a permanent head until the late summer of 1973, when Secretary Brinegar nominated James B. Gregory, another ex-Union Oil Co. manager, to run the agency. Like Brinegar, Gregory lacked experience in auto safety matters—though Gregory did have impressive scientific credentials as a research chemist and possessed some knowledge of the auto industry.

As a prelude to his nomination, Gregory met with Joan Claybrook, who was then an aide to Ralph Nader. As Gregory recalls the meeting, "it was sort of a job interview since the White House was looking for someone who Nader would find tolerable."[83] In this meeting, Claybrook made Nader's two priorities clear to Gregory: mandatory passive restraints and recall of defective vehicles. Claybrook recalls that "Gregory was a kind man—a sort of passive technocrat who seemed a bit insecure."[84] His confirmation hearings were not controversial.

Brinegar told Gregory that he wanted him at NHTSA because the agency was in a "shambles" and it needed a manager with strong technical skills. Gregory explains: "Brinegar told me to go to it, I'm here if you need me."[85] Since Brinegar and Gregory had known each other at Union Oil, they had little trouble establishing a good working relationship.

Congress and Belt Use Laws

While America was flirting with the interlock device, the rest of the developed world was enacting mandatory belt use laws. Such laws in Australia—the first in the world—proved to be an unqualified success in terms of both safety and public acceptance and were soon followed by similar laws in New Zealand, France, and Czechoslovakia. The good news became known quickly within the American safety community.

In 1972 and 1973, an associate minority counsel to the House Committee on Public Works, Richard Peet, worked with Representative William Harsha (R–Ohio)—the committee's ranking minority member—to establish a program to encourage states in America to follow Australia's lead. Peet was optimistic that a sustained campaign for belt use laws could achieve good results in the United States.

Peet had emerged as Congressman Harsha's point man on highway safety issues and was apparently responsible for drafting much of the 1973 Highway Safety Act, legislation which amends the 1966 Safety Act. In the late 1960s Harsha's son had become temporarily paralyzed in a crash. The personal experience motivated Harsha and Peet to improve pavement markings on highways, to remove roadside ("killer") boobytraps, and to stop unsafe highway maintenance and construction practices. Peet ultimately left the Hill in 1973 to create and lead a new interest group called Citizens for Highway Safety.[86]

Harsha and Peet saw that both the Nixon administration and the Democratic leadership in Congress were not taking the initiative on seat belt laws. They therefore moved into the legislative vacuum with the quiet support of both seat belt makers (especially Charles Pulley of Irvin Industries) and lobbyists from the motor vehicle manufacturers.

Their approach was to seduce state legislators with highway funds. Section 219 of the Highway Safety Act of 1973 provided a bonus of 25 percent of federal highway monies to each state that enacted a compulsory seat belt use law. Another 25 percent incentive grant was available to those states that achieved the most significant reductions in the highway death rate. According to Peet, "the first incentive was designed to encourage states to adopt seat belt use laws; the second to encourage them to undertake vigorous campaigns to enforce such laws."[87]

Within six months of passage of the 1973 act, thirty-two state legislatures were considering mandatory seat belt use legislation. By the end of 1973 mandatory seat belt laws seemed to be at least as likely to become reality as passive restraints. Puerto Rico adopted a belt use law in late 1973, and Congress authorized $25 million of highway incentive grants in 1974 for states that would enact belt use laws.[88]

NHTSA Administrator James Gregory thought Congressman Harsha's approach was a good one and devoted some personal energy to the cause. The strategy, says Gregory, was "to get one big state to go for it and then hope for a domino effect."[89] Unfortunately, says Gregory, "I was given the cold shoulder on seat belt laws by Governor Ronald Reagan's office in my home state of California. We had some support in New York State," Gregory recalls, "but we came closer in the states of Oregon and Wisconsin where I had earnest conversations with state legislative leaders in the final days before votes were taken."[90]

In December 1973 the Department of Transportation sponsored a three-day workshop on mandatory seat belt laws as a vehicle to communicate the availability of incentive grants and to publicize the success of belt use laws in Australia. Gregory made the concluding conference address. The level of interest in belt use laws among state legislators accelerated rapidly after the conference.[91]

One of Gregory's great frustrations during this period was that the so-called safety advocates, especially Ralph Nader and Joan Claybrook, were not enthusiastic about Harsha's effort. Gregory laments: "If we had gained Nader's support, we would have won."[92] Instead, state legislators were still deliberating as cars with interlocks began to hit the streets in large numbers.

Gregory's Proposal

The indecisiveness of NHTSA on the issue of mandatory passive restraints was a factor in product planning decisions at General Motors. Although NHTSA was permitting GM to install air bags (instead of interlocks) for two years with compliance tests based on Hybrid II, this permission was scheduled to expire August 15, 1975. The uncertainty about NHTSA's future requirements strengthened the argument for prolonged use of interlocks. There was particular interest in Detroit about consumer reaction to the interlock, in that former Administrator Toms had set 85 percent compliance as the target level that might justify elimination of the passive-restraint requirements. Gregory was in the process of supervising his own cost-benefit analysis of interlocks versus passive restraints.

President Cole of GM was confronting a $200 million investment in tooling and facilities for air bag production at a time of uncer-

tainty about whether the NHTSA standard would be postponed
and what certification procedures would be included with any future standards.[93] David Martin of GM recalls being told by top
management to "get the dummy work done for NHTSA so we can
move forward on air bags."[94] An October 1973 petition to NHTSA
from Volvo echoed Cole's concerns, stating that uncertainty about
the revised standard "has kept us from making crucial decisions in
regard to these [occupant-restraint] programs, and, in many cases,
has made it necessary for us to commit considerable resources to
program areas that may ultimately be discarded."[95] Administrator
Gregory insisted that he was "reviewing the issue on the merits."[96]

Gregory's investigation was slowed when Secretary Brinegar was
informed by aides in the Secretary's office that some career
NHTSA officials were "fudging the data" to support a new passive-
restraint rule.[97] Brinegar ordered his staff in DOT to conduct an
independent study of the issue, which ultimately took almost a year
to complete. The study came to no earth-shattering conclusions,
and the issue went back to Gregory for resolution.

In late March 1974 Gregory proposed that passive restraints be
required for 1977 and later models, an extension of one year beyond that contained in the rule overturned by the Sixth Circuit in
December 1972.[98] As Gregory explains, "I wanted to get the
rulemaking back on track."[99] The proposal allowed manufacturers
to choose interlocks or passive restraints in 1974–1976 models.

In support of the proposal, NHTSA cited public complaints
about the interlock and tallies of belt use for 1974 models that were
"running at or below the 60 percent level."[100] In addition to endorsing air bags, the Gregory proposal stated that passive belt systems,
such as the design developed by Volkswagen, "show promise of
achieving close to 100% usage."[101]

Reactions to Gregory's proposal were bipolar and vocal. The
Center for Auto Safety endorsed Gregory's "courageous decision on
passive restraints."[102] Allstate Insurance Company and other insurers concurred.[103] Gregory recalls that he received a personal letter
from John Volpe congratulating him for the ambitious proposal.[104]
But the top management at both Ford and Chrysler strongly opposed the proposal.[105]

In two meetings with Secretary Brinegar and Administrator
Gregory the leaders from Ford (Iacocca) and Chrysler (Newberg)
presented their case against passive restraints. As Gregory recalls
the meetings, "they basically cried economics: 'We cannot bear
these costs.'"[106] GM remained supportive of the air bag technology, but denounced the Gregory proposal—especially the 1977

deadline.[107] Industry leaders made clear that another lawsuit was coming if Gregory went forward with his rule-making proposal.[108]

As an economist, Secretary Brinegar was sympathetic with the need for cost-benefit analysis of new regulations on industry. Despite his support for air bag technology, Brinegar knew that the energy crisis was damaging the economic strength of Detroit. Bob Carter of NHTSA recalls Brinegar's sentiment that "we cannot burden this industry with such an ambitious regulation at this time."[109] Yet Brinegar's public reservations were directed primarily at the challenge of mass producing air bags with the high degree of reliability required.

Meanwhile, all of the domestic vehicle manufacturers were endorsing mandatory seat belt use laws as an alternative to mandatory passive restraints. Bills calling for such laws were introduced in twenty states by August 1974, though none were enacted into law. The campaign came very close in five states where bills passed one house of the legislature but not the other.[110] Before any bills were enacted, NHTSA's permissive policy toward interlocks blew up in its face and choked off the momentum behind both passive restraints and belt use legislation.

The Interlock Fiasco

As new cars with interlocks were sold to motorists in late 1973 and early 1974, manufacturers began to receive complaints. Reports of mechanical failures and consumer irritation with the device were numerous. The belt systems in Chrysler's cars were very poorly designed. Members of Congress and NHTSA also received a modest amount of mail from extremely annoyed constituents.[111]

Within a year, NHTSA's occupant-restraint program became the subject of legislative debate and action. A final decision on Gregory's passive-restraint proposal was shelved due to a large number of critical comments and emerging congressional backlash against the interlock device.

Richard Peet described the public reaction to interlocks:

> *The [interlock] system epitomized what I call the nuisance/nag approach to highway safety. It nagged you with buzzers and prevented you from starting your car until you buckled up. In its own irritating way, interlock worked. But it drove motorists—including Congressmen—wild.*

A wave of anti-interlock fever swept the Congress.[112] The irate letters to congressmen came at the same time that America appeared to be growing weary of regulation. Ralph Nader's political appeal was waning, while economic concerns turned politicians on to a sort of an antiregulation kick.

Congress ultimately decided to undo the interlock requirement. The Motor Vehicle and Schoolbus Safety Amendments of 1974, as proposed by the House Committee on Oversight and Investigations, amended NHTSA's occupant-restraint rule to allow consumers to choose either interlocks or a sequential reminder system (i.e., a system that would activate a buzzer and warning lights if belts were not fastened).

During debate on the House floor, Representative Louis C. Wyman (R–N.H.) offered a stronger substitute amendment. Cars could be equipped with integrated shoulder and lap belts coupled with a dashboard warning light. The Wyman amendment also prohibited DOT from mandating interlocks, sequential buzzers, or passive restraints on future models unless such devices were optional to the purchaser of a car. On the strength of anti-interlock sentiments, this amendment passed the House on a resounding vote of 339 to 49.[113]

The wide support for Wyman's amendment reflected more than just a protest against the interlock device. His amendment was symbolic of the public's growing dissatisfaction with social engineering by government—so-called Big Brotherism. A vote on behalf of Wyman was in effect a vote to rein in excessive and intrusive bureaucrats.

It is not clear whether most members of Congress understood that the Wyman amendment also included the prohibition of mandatory passive restraints. Air bag advocates were shocked. If it had been enacted by the entire Congress, the Wyman amendment would have precluded adoption of Gregory's March 1974 proposal of mandatory passive restraints.

The Senate version of the Motor Vehicle and Schoolbus Safety Amendments was passed in 1973, before the emergence of anti-interlock sentiments. Since the Senate version contained no provision about occupant restraints, the fate of interlocks and passive restraints had to be resolved by a House-Senate conference committee. Air bag advocates were optimistic because the Senate Commerce Committee Chairman, Warren Magnuson (D–Wash.), was a staunch supporter of mandatory passive restraints and a senior member of the conference committee.

Before the conferees met, Senators James Buckley (R–N.Y.) and Thomas Eagleton (D–Mo.) announced publicly their plans to endorse a prohibition of DOT's interlock rule. To place the entire Senate on record against interlocks, Senators Buckley and Eagleton sponsored an amendment to a highway bill that would have outlawed interlocks and required public hearings and congressional approval before DOT could issue any other mandatory occupant-restraint standard, including passive restraints. Although the amendment passed the Senate on a vote of 64 to 21, Senator Buckley was later compelled to withdraw the amendment since the matter was under the oversight authority of Magnuson's Commerce Committee, not the Public Works Committee. Buckley emphasized nonetheless that the roll-call vote should be viewed as an unofficial instruction to Senate conferees to accede to the House position on interlocks.[114]

On October 8, 1974, House-Senate conferees filed their report after extensive negotiations. It called for a prohibition of DOT's authority to mandate sequential warning devices with lights and buzzers as well as seat belt interlocks. DOT was permitted to require integrated lap/shoulder belts with an eight-second buzzer warning when belts were unfastened. The conferees—at Magnuson's insistence—also permitted DOT to pursue a passive-restraint rule, but DOT was required to hold a public hearing and Congress would have sixty days to disapprove the rule by concurrent resolution (i.e., a two-house legislative veto).[115]

The House and Senate passed the conference report in early October 1974 by voice vote and with little debate. This legislation spelled the death of the interlock rule and created considerable uncertainty about the future of passive restraints.

The irony of the interlock episode is that the technology was an effective use-promoting device. Surveys by NHTSA found belt use rates as high as 75 percent in some cars with the interlock feature.[116] NHTSA officials, however, never made a concerted effort to defend the technology. The prevailing view at the agency was that the interlock "was not invented here."[117] Even if such a campaign had been mounted, it is not clear that it could have been executed effectively. The leadership of NHTSA was strong on technical expertise but weak on public relations skills.

Interlock Fallout

The federal policy of providing states with incentive grants to encourage seat belt laws was also sabotaged during the period of anti-interlock sentiments. Without hearings or advance warning, an amendment sponsored by Representative Silvio Conte (R–Mass.)

was adopted by the House Appropriations Committee that prohibited the expenditure of any funds under the Section 219 highway program. Conte was the ranking minority member of the Appropriations Committee and a liberal Republican from the rural areas of western Massachusetts. He was opposed to belt use laws and strongly opposed to what he called an attempt by the federal government to "bribe" state governments. Congressman Harsha lost a floor fight over the amendment after a debate against Representative Bud Shuster (R–Pa.), a conservative member who objected to belt use laws as an intrusion on personal rights. [118]

The Senate refused to adopt the funding restriction in conference, and the offending amendment was ultimately stricken from DOT's appropriation bill. Members of the Senate Appropriations Committee were still supportive of the Section 219 program, but no funds were authorized for the incentive grants. The congressional controversy caused NHTSA to abandon efforts to provide financial incentives to states that enacted belt use legislation. As Gregory recalls it, "We drew in our horns due to the public furor." [119] According to Richard Peet, "once the states got wind of [Gregory's position], the momentum to approve belt-use laws collapsed." [120]

Conclusion

The flaw in the technology-forcing strategy employed by Volpe and Toms is that it generated so much corporate resistance at Ford and Chrysler that NHTSA lost control of the issue to other power centers (the White House and the Sixth Circuit). As political appointees in a Republican administration with close ties to corporate America, Volpe and Toms did not possess the political clout required to implement their preferred strategy, even though they were under pressure from segments of Congress to push harder. By losing control of the issue, Volpe and Toms permitted Iacocca to manipulate NHTSA's authority in the interests of Ford Motor Company.

Although air bag advocates dislike Iacocca because he squashed their technology, a strong case can be made that Iacocca's tactics were in the strategic interests of Ford Motor Company. Ed Cole at GM had left Ford in the dust in air bag development and Iacocca was surely correct in his judgment that Ford could not match GM's technical resources in an air bag development battle. The interlock was an inexpensive alternative to the air bag that proved to be an effective use-promoting device.

Although the interlock ultimately proved to be politically unacceptable, there is little evidence that Ford was harmed by this outcome. To the contrary, the interlock fiasco actually served to undermine NHTSA's political legitimacy by poisoning popular support for safety regulation. Since Gregory chose not to defend the interlock technology, the agency was placed in a very embarrassing position. Given that GM was better equipped to respond to "technology-forcing" regulation than Ford, any weakening of NHTSA would reap competitive benefits for Ford.

The adverse fallout from the interlock fiasco was widespread. Congressman Harsha's creative incentive-grant program for belt use laws was sabatoged by anti-interlock fever. Gregory's efforts to propose a mandatory passive-restraint standard also were delayed. Any future effort to promote air bags was complicated by the congressional furor surrounding the interlock. By the end of 1974, there was no strategy at NHTSA to bring passive restraints to the marketplace or to increase the rate of manual belt use.

Endnotes

1. Interview of Roy Haeusler, September 16, 1987.
2. Interview of Don McGuigan, September 3, 1987.
3. Lee Iacocca, *Iacocca: An Autobiography* (Toronto: Bantam Books, 1984), p. 296.
4. Iacocca, pp. 296–297; Joel Eastman, *Styling Versus Safety* (New York: University Press of America, 1984), pp. 228–233; "Forget 1956: Ford Stresses Safety in New Ad Campaign," *Automotive News* (December 1, 1986), p. 14.
5. Ralph Nader, *Unsafe at Any Speed* (New York: Grossman Publishers, 1972); Brian O'Neill, "Vehicle Safety: A Marketable Commodity," Paper presented at National Association of Independent Insurers, 28th Annual Workshop, April 1982, Phoenix, Arizona.
6. Iacocca, p. 297.
7. Ibid., p. 293.
8. Ibid., p. 298.
9. "Ignition Interlock: Ford's Better Idea," *Status Report: Highway Loss Reduction* (September 9, 1974), p. 2.
10. Ibid.
11. *Federal Register*, 36:8,296 (1971).
12. Interview of Douglas Toms, July 20, 1987.
13. Helen Kahn, "Tape Tells of Ford Pitch to Nixon," *Automotive News* (December 6, 1982), pp. 2+; "Secret '71 Meeting Hindered Car Rule," *Pittsburgh Press* (November 29, 1982), p. A4.
14. Interview of Robert Carter, August 18, 1987.
15. Ralph Nader Interview with Douglas Toms, January 5, 1976, mimeo.
16. Ibid.
17. Interview of John Volpe, CBS "60 Minutes" Program, 1985.

18. Interview of Douglas Toms, July 20, 1987.
19. Ibid.
20. Ibid.
21. *Federal Register*, 36:19, 266 (1971).
22. Ibid.
23. "DOT Gives Two-Year Delay on Air Bags," *Status Report: Highway Loss Reduction* (October 4, 1971), p. 3.
24. *Federal Register*, 37:3, 911 (1972).
25. Interview of Douglas Toms, August 18, 1987.
26. Interview of David Martin, September 3, 1987.
27. "Blankets and Nets and Court Action?" *Automotive Industries* (November 1, 1971), p. 22.
28. "Air Bag Testing Lags as Detroit Races Deadline," *Industry Week* (March 6, 1972), pp. 81–83.
29. Interview of Brian O'Neill, March 3, 1988.
30. Ibid.
31. Interview of Benjamin Kelley, July 15, 1987.
32. "Haddon Calls for Air Bag Field Tests," *Status Report: Highway Loss Reduction* (October 4, 1971), p. 4.
33. Ibid.
34. *Status Report: Highway Loss Reduction* (August 21, 1972), p. 4.
35. Interview of Douglas Toms, July 20, 1987.
36. Ibid.
37. Ibid.
38. *Cumulative Regulatory Effects on the Costs of Automotive Transportation*, White House Office of Science and Technology Policy, Washington, D.C., 1972.
39. "A Backlash Against New-Car Standards," *Business Week* (March 25), 1972, p. 23.
40. Interview of Lowell Dodge, August 13, 1987.
41. "A Backlash Against New-Car Standards," p. 23; Ralph Nader, "Washington Under the Influence: A Ten-Year Review of Auto Safety Amidst Industrial Opposition," 1975. Unpublished memorandum.
42. Written statement of Dr. Lawrence A. Goldmuntz, in *Regulatory Reform— Volume IV*, Hearings Before Committee on Interstate and Foreign Commerce, U.S. House of Representatives, 94th Congress, 2nd Session, 1976, pp. 439–449.
43. "A Jarring Letdown for the Air Bag," *Business Week* (April 29), 1972, p. 19.
44. "Bag Tests Are Questioned," *Automotive Industries* (August 15), 1972, p 65; "Air Bag Test Failure Laid to Obsolete Parts," *Status Report: Highway Loss Reduction* (July 17, 1972), p. 1.
45. Ibid.
46. Interview of James Hofferberth, December 14, 1987.
47. G.W. Trinca and B.J. Dooley, "The Effects of Seat Belt Legislation on Road Traffic Fatalities," *Australian-New Zealand Journal of Surgery* 47:150–55 (April 1977); F.T. McDermott and D.E. Hough, "Reduction in Road Fatalities and Injuries after Legislation for Compulsory Wearing of Seat Belts:

Experience in Victoria and the Rest of Australia," *British Journal of Surgery* 66:518–521 (1979); "Australian Experiment Found Successful," *Status Report: Highway Loss Reduction* (June 12, 1972), p. 3.

48. "Belt Bandwagon Gaining Riders," *Status Report: Highway Loss Reduction* (June 12, 1972), pp. 6–8; "Court Delays Auto Air Bags: Calls for More Refined Testing," *Industry Week* (December 11, 1972), p. 20.
49. Interview of James Hofferberth, December 14, 1987.
50. *Federal Register*, 36:12,866 (1971).
51. Joseph M. Callahan, "Belt Systems Acceptable for 1976 Doug Toms Reveals in AI Interview," *Automotive Industries* (June 15, 1972), p. 18.
52. "Will VW's New Belts Replace Air Bags?" *Industry Week* (May 29, 1972), p. 73; "VW on Safety: Air Bag, Nein; Gas Belt, Ya," *Iron Age* (May 25, 1972), p. 27.
53. Interview of Robert Carter, August 17, 1987.
54. "Belt Systems Acceptable for 1976 Doug Toms Reveals in AI Interview," p. 18.
55. William Bowen, "Auto Safety Needs a New Road Map," *Fortune* (April 1972), pp. 98–101+.
56. "Air Bag Problems Still Ballooning," *Industry Week* (May 1, 1972), p. 65.
57. *Chrysler Corporation* vs. *Dept. of Transportation*, 472 F.2d (6th Circuit 1972), pp. 659–81.
58. Interview of Lowell Dodge, August 13, 1987.
59. *Chrysler* vs. *DOT*, p. 659.
60. Interview of Lowell Dodge, August 13, 1987.
61. Interview of Don McGuigan, September 3, 1987.
62. Interview of Ralph Hitchcock, July 23, 1987.
63. Ibid.
64. Interview of Roy Haeusler, September 16, 1987.
65. Interview of Milford Bennet, September 3, 1987.
66. Interview of Roger Maugh, July 8, 1987; interview of Don McGuigan, September 3, 1987.
67. "GM Offers Air Bag; Others Use Inertia Belt," *Industry Week* (February 19, 1973), p. 7; "Air Bag Test Fleet Program Faltering," *Status Report: Highway Loss Reduction* (August 21, 1972), pp. 4–5.
68. "Air Bag Tests Now Use People," *Industry Week* (January 29, 1973), p. 73.
69. *Status Report: Highway Loss Reduction* (April 9, 1973), p. 1.
70. Interview of David Martin, September 3, 1987; *Federal Register*, 38:8,455 (1973).
71. "GM's Air Bag Program Stalled," *Automotive Industries* (July 15, 1973), pp. 18–19.
72. Richard A. Wilson and Carl M. Sagave, "Restraint System Effectiveness—A Study of Fatal Accidents" (presented at the GM Safety Seminar, Warren, Michigan, June 20–21, 1973).
73. "GM's Air Bag Program Stalled," p. 19; "GM Shifts Air Bag Policy," *Status Report: Highway Loss Reduction* (July 10, 1973), pp. 1–2.
74. Interview of David Martin, September 3, 1987.
75. Ibid.

76. "Now Down to Real World Testing," *Automotive Industries* (September 1, 1973), p. 15; "NHTSA Issues New Air Bag Test Specifications," *Status Report: Highway Loss Reduction* (April 9, 1973), p. 1.
77. "Auto Safety Might Sell, Consumer Survey Finds," *Industry Week* (October 8, 1973), p. 73; "GM Slashing Air Bag Plans," *Status Report: Highway Loss Reduction* (October 17, 1973), pp. 7–8.
78. "Pressure Mounts on NHTSA to Drop 'No Start' Rule," *Status Report: Highway Loss Reduction* (April 9, 1973), pp. 2–3.
79. Ibid.
80. "NHTSA Keeps 'No Start' Rule for 1974 Models," *Status Report: Highway Loss Reduction* (April 24, 1973), p. 8.
81. "Ignition Interlock: Ford's Better Idea," *Status Report: Highway Loss Reduction* (September 9, 1974), p. 5.
82. Interview with James Gregory, July 14, 1987.
83. Ibid.
84. Interview of Joan Claybrook, July 24, 1987.
85. Interview of James Gregory, July 14, 1987.
86. Testimony of Richard Peet, President, Citizens for Highway Safety, DOT's Public Hearings on Passive Restraints, Washington, D.C., December 7, 1983, pp. 1–15.
87. Ibid.
88. Ibid.
89. Interview of James Gregory, July 14, 1987.
90. Ibid.
91. "DOT Pushing Belt Use Laws," *Status Report: Highway Loss Reduction* (December 20, 1973), pp. 5–6.
92. Interview of James Gregory, July 14, 1987.
93. "DOT Still Indecisive on Passives," *Status Report: Highway Loss Reduction* (February 21, 1974), pp. 1–2; "GM Puts Air Bag Back in Limbo," *Automotive Industries* (February 15, 1974), p. 19.
94. Interview of David Martin, September 3, 1987.
95. "DOT Still Indecisive on Passives," p. 3.
96. Ibid., p. 1.
97. Interview of Robert Carter, August 17, 1987.
98. *Federal Register*, 39:10,271 (1974).
99. Interview of James Gregory, July 14, 1987.
100. "NHTSA Issues Passive Restraint Proposal," *Status Report: Highway Loss Reduction* (March 26, 1974), pp. 1–3.
101. Ibid., p. 3.
102. Ibid., p. 3.
103. "Insurers Supporting Passive Restraints," *Status Report: Highway Loss Reduction* (May 23, 1974), p. 1.
104. Interview of James Gregory, July 14, 1987.
105. "Auto Makers Renew Passive Restraint Attack," *Status Report: Highway Loss Reduction* (June 18, 1974), p. 1.
106. Interview of James Gregory, July 14, 1987.
107. "Air Bags Head for the Courts," *Business Week* (July 6, 1974), pp. 50–51.

108. Ibid., p. 51; "Auto Belts, Bags, Buzzers on Safety Collision Course," *Iron Age* (August 19, 1974), pp. 46–47.
109. Interview of Robert Carter, August 17, 1987.
110. "Auto Belts, Bags, Buzzers on Safety Collision Course," p. 47.
111. "Mandatory Seat Belt System Voided," *Congressional Quarterly Almanac* (1974), pp. 685–688.
112. Testimony of Richard Peet, President, Citizens for Highway Safety, DOT Hearings on Passive Restraints, Washington, D.C., December 7, 1983, p. 4.
113. "Mandatory Seat Belt System Voided," p. 687.
114. Ibid., p. 688.
115. Ibid.
116. *Safety Belt Interlock System Usage Survey,* U.S. Department of Transportation, NHTSA, DOT-HS-801-594 (1976), p. 1.
117. Interview of Michael Finkelstein, December 9, 1987.
118. "House Would Cut Funds for Belt Law Incentives," *Status Report: Highway Loss Reduction* (July 8, 1974), p. 12.
119. Interview of James Gregory, July 14, 1987.
120. Testimony of Richard Peet, p. 4.

Chapter 5

NEGOTIATING SAFETY

Advocates of pluralism envision the public interest evolving out of a blending of the preferences espoused by lobbying groups. Just as classical economists believe that perfect market competition achieves optimal prices and production via the "invisible hand," democratic pluralists believe the public interest can be achieved without any central planning or blueprint. Some neo-pluralists argue that interest group competition over regulation can be harnessed for the public good, but only if it is managed through some formal negotiation process.

Formal bargaining has been proposed as a complement to or substitute for the normal rule-making processes of federal agencies. One version of this approach places the administrative agency in the role of mediator while corporations and consumer activists negotiate a rule that both parties can tolerate. Another version of formal bargaining has the regulator on one side of the table and regulatees on the other. The latter approach was used by the Department of Transportation in 1976 in an effort to resolve the air bag controversy. A highly skilled administrator came away with a remarkable agreement that promised to deliver a program that was urgently needed: a large-scale field test of air bags.

Enter William Coleman

Claude Brinegar resigned as secretary of transportation in early 1975 after roughly two years of service. He had taken some personal interest in the passive-restraint issue but did little to resolve the controversy. To replace him President Ford appointed William T. Coleman, Jr., a black attorney and liberal Republican.

Coleman's friends and foes describe him as an intriguing personality—a shrewd public official with a strategic legal mind, a substantial ego, and a delightful sense of humor. He also had a reputation for being a stickler for due process and public participation in governmental decision making.[1] Consumer activists and industrial lobbyists were not sure what to expect from Coleman on air bags since he was known more for his civil rights work than for his expertise in transportation safety policy.

Coleman elected to retain Gregory at the helm of NHTSA. This was a significant decision because, like his two predecessors in the job (Haddon and Toms), Gregory had become very sympathetic with the case for passive restraints. As Coleman took office, Gregory was organizing public hearings on his proposal of March 1974 to make passive restraints mandatory beginning with the 1977 model year.

Over fifty witnesses testified at the five days of hearings in May 1975. Gregory found that the interested parties were more sharply polarized on the issue than every before. Ralph Nader, consumer groups, insurance industry representatives, and public health groups supported Gregory's proposal, while the entire motor vehicle manufacturing industry, including General Motors, opposed mandatory passive restraints in general and air bags in particular.[2] The new position of GM was a major blow to air bag proponents, who were left with no allies in the industry.

GM's New Strategy

At the insistence of GM President Ed Cole, front-seat air bags were offered as a $225 to $315 option on several of GM's luxury car lines during the 1974–1976 model years. GM reported spending $60 million to acquire a capacity to produce 100,000 air bag–equipped cars per year, but only about 10,000 cars were actually built and sold with the devices.[3] The lack of sales was remarkable considering that consumers who elected air bags in model year 1974 did not have to deal with the unpopular starter-interlock system.

One GM official noted that the air bag option was continued through the 1976 model year only because "we still have a little material left."[4] By the time GM officials testified before Gregory in the spring of 1975, the company had abandoned plans to offer air bags on 1977 and later models. Indeed, the lines on which the air bags were offered went out of production and were replaced by downsized cars.[5]

Although the air bag option was introduced late in the 1974

model year, some efforts were made by GM to market the option. Dealers were provided with movies of air bag crash performance to show to prospective customers. Eighteen city newspapers and several national news magazines were targeted with advertisements promoting the air bag.[6] Despite these efforts, subsequent surveys of 1974 GM buyers revealed that some customers were aware of the availability of the air bag option while others were not. Few of GM's 6,500 dealers embraced the "Air Cushion Restraint System" as a means of increasing sales.[7]

To understand why GM's air bag program failed, NHTSA hired Booz, Allen and Hamilton, Inc. to interview GM management, dealers, and customers. The firm concluded that the three key problems were lack of dealer commitment, unanticipated consumer concerns about the device's safety, and availability on large cars instead of small cars (even though owners of small cars might have been more receptive). They also found that GM's top management turned against the device even before the marketing campaign had been fully implemented.[8]

A reporter for the *Wall Street Journal*, Albert Karr, investigated GM's optional air bag program in early 1976. His front-page story also concluded—based on a survey of GM dealers, car buyers, and industry observers—that "the air bag received no wholehearted promotion."[9] It seems that many dealers knew little about the air bag and did little to whet consumer interest in the device. Retired GM President Ed Cole insisted in an interview with Karr that an option like the air bag can sell only if a company "creates a desire on the part of the user to buy it."[10] Cole told Karr that GM never made the effort to create this desire, a claim that was disputed by William Buxton, GM's sales vice president. According to Buxton, "we gave it one hell of a try," yet buyers still did not want the air bag.[11]

GM's optional air bag program faltered at the same time as the leadership of GM soured on the program. This is a case where cause and effect are difficult to identify. The change in managerial policy has been attributed to the apparent difficulty in selling cars with air bags. GM's David Martin argued that there was "no consumer demand for them."[12] Marketing and financial people at GM also saw the air bag as a loser. Since NHTSA's rule was not in effect, the argument for air bags within GM had weakened.

Skeptical GM lawyers also pointed to the high ratio of lawsuits to air bag deployments. From a technical point of view, GM engineers found that the device performed on the road as designed, yet GM lawyers reported fourteen lawsuits involving air bags after only sixty deployments.[13] These numbers confirmed earlier suspicions of auto executives that air bags might prove to be a liability nightmare.

The company's move away from the air bag was assured when GM President Ed Cole retired in September 1974. Although GM's Air Cushion Restraint System had not proved to be as promising as Cole had originally hoped, he was still enthusiastic about the idea. David Martin recalls that Cole was an extremely optimistic and ambitious leader who did not like hearing about the lingering technical difficulties that troubled working-level engineers.[14] Cole's retirement left the air bag with no leader at the corporation committed to making the air bag program a success. The air bag issue fell to the hands of Executive Vice President Roger Smith and Chairman Richard Gerstenberg, who both saw the device as part of GM's larger regulatory problem.

Detroit's Economic Woes

The Arab oil embargo of October 1973 and the quadrupling of oil prices by the OPEC cartel produced a severe worldwide recession in 1974 and 1975. Domestic car sales plummeted from 9.7 million units in 1973, to 7.5 million units in 1974, to 7.1 million units in 1975. Detroit simultaneously confronted intensified competition from low-priced Japanese imports. Long gas lines and increases in the price of gasoline caused many customers to shift to these smaller, more fuel-efficient imports. From 1973 to 1975 the Japanese share of the new-car market increased from 6.5 to 9.4 percent. The historically profitable Big Four soon found their financial positions deteriorating and in some cases sinking into the red.[15]

Detroit turned to Washington for help. Economists had estimated that federal regulations of emissions and safety were increasing significantly the cost of producing a new car.[16] As a result, Detroit, led by Gerstenberg, was lobbying Congress and the Ford administration for a five-year moratorium on new auto regulations of any kind.[17]

Congressional leaders were divided on how Gregory should react to Detroit's proposal to halt new regulations. Senator Thomas Eagleton (D–Mo.) expressed a prevalent view: "Congress would be very reluctant to endorse anything that would further raise the price of cars. There would be support for further testing, but not a mandatory standard for the next couple of years."[18] In opposition to Senator Eagleton were the long-standing advocates of passive restraints, including Senators Warren Magnuson (D–Wash.), Vance Hartke (D–Ind.), Frank Moss (D–Utah), and Congressmen Harley Staggers (D–W. Va.) and John Moss (D–Calif.). In a jointly signed letter to Administrator Gregory, these legislators urged Gregory to resist Detroit's call for a "misguided moratorium" on new safety regulation.[19]

Debating Benefits and Costs

Although Administrator Gregory was concerned about inflation and the energy crisis, he insisted that "the costs of not going to improved safety systems in terms of lives and injuries cannot be ignored."[20] At the 18th Stapp Car Crash Conference, Gregory presented the results of his staff's cost-benefit analysis of the mandatory passive-restraint proposal. Even after revising an earlier draft of the study in response to comments from manufacturers, the final draft concluded that the air cushion/lap belt system had a favorable cost-benefit ratio. Air bag installation in all new cars was projected to prevent 11,600 fatalities per year when all cars on the road were equipped with the device. Gregory recalls that "regulation was in such disfavor at this time that even a rule with a favorable cost-benefit ratio was viewed skeptically by many politicians."[21]

Another cost-benefit study prepared for Allstate Insurance Company by former GM Vice President John DeLorean reached even more favorable conclusions about the air bag. DeLorean concluded in public testimony before NHTSA that the trend to smaller cars would result in a 40 percent increase in highway deaths and injuries from 1980 to 1990. He estimated that delaying the introduction of front-seat air bags as standard equipment for only three years would result in 37,000 needless deaths and more than $18.6 billion in societal loss due to injuries and fatalities.[22]

The emergence of the insurance industry as a vocal supporter of the passive-restraint proposal added a new economic dimension to the issue. Allstate was offering a 30 percent discount on medical and no-fault personal injury coverages for air bag–equipped cars. Other insurers had followed Allstate's lead. Allstate also purchased for its fleet the last 200 air bag–equipped 1976 Oldsmobiles manufactured by General Motors before it closed its air bag assembly line.[23] At Gregory's May 1975 hearings the insurance industry representatives, such as Allstate Vice President Donald Schaffer and Nationwide's Douglas Fergusson, were among the most well-informed participants. They argued in favor of mandatory passive restraints. Their motives were hardly altruistic considering they may have expected passive restraints to reduce their company's payments to policyholders involved in crashes.

Administrator Gregory was not precisely sure what next steps were appropriate. As he recalls this period, "the interlock fiasco lingered over the entire issue and thus we had basically given up on mandatory seat belt wearing laws."[24] Gregory described driver-side air bags as "ready to go on a large-scale basis," but was worried that "reliability of the technology might be difficult to sustain in mass production."[25] Reliability was more than a technical and liability concern; it also was a public acceptance issue. Gregory explains: "Many citizens were suspicious of NHTSA generally and frightened about driving with an explosive device in their car."[26]

A Suggestion from White House Economists

In response to inflationary pressures in the economy, President Gerald Ford issued in the fall of 1974 a requirement that all major federal regulations be subjected to "Inflationary Impact Statements."[27] This requirement was issued just prior to the arrival of George Eads at the Council on Wage and Price Stability (COWPS), the White House office charged with the responsibility of implementing President Ford's order. As assistant director of government operations and research at COWPS, Eads and his small staff of economists geared up to do economic critiques of agency regulations.

Eads was trained at Yale in economics and was eager to use his position to enhance the quality of benefit and cost analysis used by federal regulatory agencies. One of his first assignments was to review, at NHTSA's request, a proposed air-brake standard for trucks. COWPS analysts found that although NHTSA's cost estimates seemed reasonable, the benefit estimates were poorly researched and appeared to be inflated. Eads submitted these findings to NHTSA's public docket, a move that created some irritation and embarrassment among NHTSA analysts.[28]

On the heels of this review, Eads received a call in late 1974 from Dr. Larry Goldmuntz, an engineering consultant and former White House science adviser. Goldmuntz recommended that Eads look into NHTSA's cost-benefit analysis of mandatory passive restraints. Eads agreed and hired Goldmuntz as a consultant to help his small staff learn the key technical issues associated with NHTSA's March 1974 proposal. A report by Goldmuntz was submitted to Eads, who used it along with other staff inputs in preparing testimony for submission to NHTSA.[29]

Eads concluded that there were major uncertainties about both the technical effectiveness of air bags and the public's likely reaction to widespread installation of the system in new cars. In a February 7 COWPS memorandum to NHTSA Administrator James Gregory,

Eads recommended that NHTSA consider a large-scale, real-world government test of the air bag technology.[30] He also presented this idea in March 1975 testimony before the consumer subcommittee of the Senate Commerce Committee and later in testimony before Gregory at NHTSA's May 1975 hearings.[31]

The rationale behind the Eads plan was to "get a lot of air bags out there quickly."[32] Eads argued that such a demonstration would remove the burden of experimentation from the general auto-buying public and result in "increased consumer confidence."[33] The $250 million testing plan could begin with the 1977 model year, though it would be necessary to delay for several years a mandatory passive-restraint standard. The 500,000 figure was selected to assure a statistically significant number of air bag deployments within several years. Once cars were in crashes, the government could buy them back for in-depth analysis.[34]

Gregory and his staff were "irritated" by the Eads proposal because the Council on Wage and Price Stability seemed to be "technically outside of their area."[35] The "brash young economists" at COWPS, Gregory recalls, tended to convey their views with "a tone of arrogance."[36] On the merits of the idea, Gregory wrote to Eads:[37]

> *Apart from the obviously difficult (but not impossible) administrative task of equitable selection [of cars to have air bags] and the formidable task of securing funds [for the demonstration], I have to say that NHTSA cannot be against the idea from the standpoint of increasing vehicle safety or acquiring additional data. Nevertheless, I also have to say that, due to our rulemaking posture, we could not commit ourselves to waiting for the results of such a test as the deciding factor; nor obviously can I say we would not take such results into serious consideration.*

The unenthusiastic response of Gregory to COWPS's proposal arose partly out of a perception by NHTSA staffers that COWPS was trying to shoot down mandatory passive restraints on economic grounds. NHTSA staff were not pleased that Eads had elected to hire consultant Laurence Goldmuntz, whom they considered to be an opponent of mandatory passive restraints. The letter to Eads from Gregory explained that "I am sure you recognize that we began our cost-benefit study early in 1974, long before the requirement for inflationary impact statements or the interest of the Council on Wage and Price Stability."[38]

The anxieties of NHTSA staffers about COWPS were inflamed when President Gerald Ford publicly questioned the cost-effectiveness of air bags in a May 1975 speech before the U.S. Chamber of Commerce. President Ford expressed skepticism about whether air bags have been "proven sufficiently cost-effective for us to require their installation in all cars at between $100 and $300 each." He concluded his remarks by calling for a reexamination of federal rule-making activity from the perspective of costs and benefits.[39] Although it is not clear whether President Ford's remarks on this issue were based on COWPS material, "it was common for COWPS to supply the White House with information to be embodied in Presidential statements."[40]

Pressures for a Decision

Gregory and his staff spent six months reviewing the new data and issues raised at the May 1975 hearings. There was the fuel crisis, the new nationwide 55 mph speed limit, the rapid market shift to small cars, the COWPS proposal, the DeLorean study, GM's field tests of air bags, new studies of restraint-system effectiveness, and additional foreign experience with mandatory seat belt wearing laws. Each of these factors needed to be weighed and reflected in the agency's final decision. When asked by congressional leaders—including John Moss (D–Calif.) in the House and Vance Hartke (D–Ind.) in the Senate—when a decision would be announced, Gregory responded that a decision should be forthcoming by the end of 1975.

The indecisiveness at NHTSA during this period was exacerbated by internal fighting among the agency's career technical people. The associate administrator of motor vehicle programs, Robert Carter, had been a zealous air bag advocate since the Volpe years and was urging Gregory to push forward with a tough regulatory program. Gene Mannela, the associate administrator for research and development, was more cautious about air bags and saw promise in more comfortable belt designs. Carter and Mannela also had a history of personal disagreements on other policy matters.[41]

To avoid this conflict, Gregory asked Howard Dugoff, the associate administrator for policy and planning, to prepare an objective analysis of the entire issue. Dugoff was an analyst who shied away from ideological positions. Dugoff's analysis was later used as the quantitative basis of decisions by both Secretary Coleman in 1976 and Secretary Brock Adams in 1977.

Gregory made his policy recommendations to Secretary Coleman in early 1976. Gregory's proposed plan was not made public by Coleman until a year later, but it called for (1) mandatory passive restraints on the driver side beginning with model year 1980 and full-front protection two years later and (2) withholding of federal highway funds from states that did not enact mandatory seat belt wearing laws.[42] These recommendations had the support of senior NHTSA officials.[43]

Further Election Year Delays

According to one NHTSA official, "the reaction of the Secretary's office was that Gregory's position was too radical, too unrealistic."[44] Rather than approve Gregory's plan at the start of a presidential election year when Detroit was suffering, Coleman decided to engage himself in a detailed review of the entire issue. He appointed an ad hoc task force led by his personal staff to examine the issue.

In late 1975 Senator Warren Magnuson, chairman of the Senate Commerce Committee, publicly accused DOT of "dragging its feet" on the passive-restraint issue.[45] Magnuson warned that "if the auto industry is to have adequate lead time in which to integrate the air restraint system into its vehicles, expeditious action from the Department of Transportation and the Ford Administration is needed now." After citing statistics of belt use around 25 percent and a trend to less crashworthy vehicles, Magnuson concluded it was urgent that DOT act on passive restraints.[46]

In January 1976 Congressman John Moss (D–Calif.), chairman of the House Subcommittee on Oversight and Investigations, wrote Gregory for an explanation of the "marathon series of delays" in the passive-restraint rule making.[47] Moss was a twenty-year veteran of the House who championed the causes of consumer protection, openness in government, and federal regulation of private enterprise. Lowell Dodge, who worked for Moss from 1975 to 1978, describes Moss as an "intense, energetic, and thorough" legislator who saw his role as a "scorekeeper" of federal agency performance.[48] According to Gregory, Moss is "one of the great California liberals whose commitment to auto safety was intense and sincere."[49]

To Moss's request for a definitive timetable for the passive-restraint rule making, Gregory responded that his goal was to issue a final rule prior to the traditional August recess of Congress. Given the polarization of views on the issue, Gregory emphasized the importance of backing up any rule making "by as meticulous an analysis and justification as possible."[50]

Ralph Nader was also perturbed by the long delays in the passive-restraint rule making, especially the months of delay following the May 1975 hearings. He wrote to Coleman in late December 1975 seeking an explanation and timetable.[51] It was not until four months later that Coleman responded to Nader by saying that he was seeking "a resolution as quickly as possible while giving all interested parties an opportunity to register their views for consideration."[52] In fact, Coleman's aide Michael Browne had been looking into the issue since early 1976.

In February 1976 Gregory announced his intent to resign, saying the job "had worn me out."[53] He was comfortable leaving the job in the midst of the passive-restraint deliberations because he knew that Secretary Coleman had taken personal control of the issue. At congressional hearings prior to his resignation, Gregory denied speculation that he had resigned due to frustrations over the passive-restraint controversy.[54]

Gregory's successor at NHTSA, John Snow, did not take charge until mid-1976. Snow was an attorney and economist with five years experience in transportation policy at DOT. Coleman's investigation of the passive-restraint issue was well underway before Snow was confirmed by the Senate.

During his short tenure as transportation secretary, Coleman had many complex issues on his agenda, including the future of Amtrak and the Concorde. When he took over the passive-restraint issue, Gregory had the agency "moving with a full head of steam" toward an industry-wide regulation,[55] despite resistance from automakers. Coleman was "looking for a way to break this logjam."[56]

On May 24, 1976, Secretary Coleman made a speech at the Economic Club of Detroit where he touched on the passive-restraint issue. In what air bag advocates saw as an obvious attempt to curry favor with Detroit in a presidential election year, Coleman made several comments about the significant costs and limited effectiveness of air bags. He also described some of the policy options he was considering, such as mandatory seat belt wearing laws.[57]

Coleman's public remarks irritated Donald L. Schaffer, a vice president of Allstate Insurance Company. Schaffer was the key force behind Allstate's pro–air bag posture throughout the 1970s. He was described by an admirer as "a feisty and combative lawyer who pursued this lifesaving technology with a singlemindedness unsurpassed by Naderites."[58] Schaffer was "despised" by many auto executives in Detroit who saw him as an "air bag crusader."[59] In a letter to Coleman, Schaffer questioned both the factual basis of the remarks and the secretary's legal authority to promulgate some of the options mentioned.[60]

Coleman's First Move

On June 9, 1976, Coleman announced the process he would use to reach a final decision on the passive-restraint issue. The process would begin with a published notice of proposed rule making covering five possible courses of action. A new public hearing to be moderated by Coleman was scheduled for August 3, 1976. The options to be discussed included (1) mandatory passive restraints, (2) state enactment of seat belt wearing laws, (3) a national passive-restraint demonstration program, (4) a mandatory passive-restraint option for consumers, and (5) continuation of voluntary manual belt systems. Coleman promised to issue a written decision by January 1, 1977.[61]

The *Washington Post* noted that the late date for a decision meant that Coleman was trying to avert a campaign issue by resolving the issue after the November 2 elections.[62] Coleman's announcement was disparaged by proponents of passive restraints. Congressman Moss charged that the process "substitutes procrastination for a policy of action to protect the driving public."[63] Administrator Gregory, who had announced his resignation in February 1976 and was serving until a successor was found, expressed "disappointment" about Coleman's inability to make a final decision by August.[64] Ralph Nader joined in denouncing the process while calling Coleman "an absolute coward in refusing to stand up to the automobile industry's pressures."[65]

Before proceeding with his announced process, Coleman wrote to Nader personally, saying that "I hope I can demonstrate to you that I was not and am not a coward."[66]

For the next several months, Coleman immersed himself in the issues and data. The public hearing in August 1976 revealed no startling facts or issues. The only surprise was the performance of Coleman, who according to most observers demonstrated a remarkable command of the facts and arguments. One insurance industry representative said: "Coleman surprised the hell out of everyone. Rather than take a ceremonial role at the hearings, he asked extremely tough and penetrating questions."[67] One of the low points of the hearing occurred when Ralph Nader drew a parallel between Coleman's influence on auto safety and the role of the Ku Klux Klan in civil rights.[68] Some observers felt Coleman appeared deeply offended by Nader's remark.[69]

How to Pay for a Demonstration

Coleman's ad hoc task force quickly concluded that some sort of demonstration program was the appropriate policy response. Some task force members saw a demonstration as necessary to prove the air bag's effectiveness; others saw it as a method to foster public confidence in the technology. The stumbling block was that the task force could not figure out how to organize and pay for a government demonstration program.[70]

At this stage, Secretary Coleman became involved personally and made a critical contribution. It was Coleman who suggested that it was the responsibility of the auto companies to organize and finance a demonstration program. Because their products were associated with the injuries, Coleman argued, it was their duty to pay for injury prevention efforts. Under Coleman's idea, DOT's role would be to negotiate the terms of the demonstration and monitor the performance of manufacturers. Since GM and Ford had recently rebounded into the black, he saw no reason why they shouldn't bear the costs of the demonstration.[71] In effect, Coleman's task force drew on COWPS's earlier call for a large-scale demonstration, but Coleman offered a different financing scheme. COWPS had reaffirmed its support for a demonstration in public testimony before Coleman at the August hearings.[72]

The Coleman Plan

The passive-restraint issue was not resolved by Coleman until after Gerald Ford was narrowly defeated by Jimmy Carter in the November 1976 elections. In December 1976 Secretary Coleman announced his findings and decision about the occupant-restraint issue. He concluded that passive restraints—both air bags and automatic belts—were feasible, would provide substantially increased occupant-crash protection, and could be provided to consumers at reasonable cost. He estimated that if passive restraints were installed in all automobiles, roughly 12,000 fewer crash fatalities and 100,000 fewer serious nonfatal injuries would result each year.[73]

Although impressed with the lifesaving potential of passive restraints, Coleman was concerned that the public might not accept a passive-restraint mandate. He emphasized the importance of public acceptance by noting the recent interlock fiasco, the refusal of state legislatures to pass mandatory belt use laws, and the decision of some state legislatures to repeal or weaken motorcycle helmet wearing requirements. Opinion polls also indicated that a significant minority of citizens were strongly opposed to mandatory passive restraints.[74]

Based on these findings, Coleman anticipated a large degree of public hostility—at least initially—to a new regulatory requirement that people be protected with passive restraints. To promote public acceptance of the new safety technologies, Coleman called for manufacturers to participate in a voluntary demonstration program involving 500,000 new cars during model years 1979 and 1980. Such a program, he argued, would increase public understanding of the technologies and foster long-term acceptance of the devices. In the short run the program would make passive restraints available to those who wanted them without penalizing consumers who wear manual belts and don't want to pay the added cost of passive restraints. In the long run, a demonstration might stimulate market demand for passive restraints and make regulation unnecessary.[75]

Christmas Negotiations

After his announcement, Coleman assigned aide Michael Browne the job of negotiating with domestic and foreign manufacturers to hammer out a viable demonstration program. Browne was a smart young attorney who had Coleman's complete confidence. His strategy was to make the auto companies an offer they could not refuse.

Browne made quite clear early in the negotiations that the only alternative to a "voluntary" demonstration program would be a decision by Coleman to require that all new cars built after September 1, 1979, be equipped with passive restraints. This informal use of administrative power was documented in a January 14, 1977, letter from the U.S. attorney general's office to DOT:[76]

> *The manufacturers were motivated to sign their separate contracts by a desire to avoid a mandate of air bags as standard equipment and the Secretary has actually stated to the automakers that in his opinion a mandate is the only alternative if a demonstration cannot be organized.*

Industry executives had good reason to believe that Coleman's successor at DOT—whomever President Carter might appoint—was likely to be even more inclined to regulate than Coleman, so the incentive to negotiate was powerful.

The negotiations organized by Browne had to be completed quickly. Sessions were held with each large manufacturer separately during the Christmas holiday season. The most important talks were with the negotiating teams from GM and Ford, which were led by Vice Presidents Roger Smith and Herbert Misch, respectively. A Ford official recalls that Browne was successful in his effort to keep each manufacturer unaware of what the other manufacturers were willing to do.[77]

In the beginning, it was not clear that Ford and GM would participate in a demonstration. Herbert Misch reportedly had difficulty persuading Henry Ford II and Lee Iacocca to negotiate with Coleman.[78] William Chapman of GM observed: "We really didn't want to go along with Coleman. The federal government had no financial investment. The liabilities all fell to the companies."[79]

In the process of negotiation, key concessions were made by both sides. Browne allowed Ford and GM to commit to building a minimum of only 10 percent of the production capacity announced by Coleman in December and relaxed the timetable by a year. But Browne extracted major concessions from Smith and Misch on price and marketing programs. And Browne resisted efforts by Ford to gain immunity from liability suits.[80]

One participant observed that Browne and Coleman appeared to be playing a "good cop/bad cop" routine, though this may not have been a conscious tactic. Browne took the hard-nosed positions, and when the talks bogged down, Coleman would arrive and offer new ways to be reasonable. Browne's tendency was to be quite aggressive, but he was restrained at various times by Coleman's aide Mary Graham.[81]

By the final days of the Ford administration in January 1977, Coleman announced publicly the terms of an agreement with General Motors, Ford, Volkswagen, and Mercedes. In total, these companies pledged to establish the production capacity for 500,000 cars with passive restraints in model years 1980 and 1981.[82]

General Motors pledged to establish the capacity for production of 300,000 mid-sized cars with full-front air bags and to build a minimum of 30,000 such cars over the two-year period. GM also pledged to spend between $5 and $15 million marketing the air bag, with the consumer price to be set by the General Accounting Office based on cost estimates for full-run production (about $100 per car). Ford promised to establish a production capacity to build up to 140,000 compact cars with driver-side air bags, including a $2 million marketing effort. Mercedes made a production capacity commitment of 2,250 cars with air bags, and Volkswagen agreed to equip 60,000 cars with automatic shoulder belts and knee pads.[83]

On the strength of these commitments, Coleman terminated the passive-restraint rule making on January 17, 1977—just as he was stepping down as secretary. GM's Smith and Ford's Misch recognized that any agreement with Coleman would not preclude any future DOT secretary from reopening the passive-restraint rule making. They therefore insisted that the written agreement with Coleman contain a clause that terminated their obligations to the demonstration program in the event that the passive-restraint rule making was reopened.[84]

Reactions to Coleman's Plan

The initial reactions of air bag advocates to Coleman's plan ranged from skepticism to vocal opposition. According to Michael Finkelstein, who later ran rule making at NHTSA during the Carter administration, "the Coleman plan was seriously flawed."[85] Finkelstein argues that if Coleman's plan had taken effect, it would not have worked because consumers would not have purchased the air bags as an option. The money earmarked for advertising was not very much in the context of auto marketing, and Coleman's plan did not assure that the devices would be offered as standard equipment.

Claybrook also emphasizes that the auto companies were not committed under the plan to produce 500,000 cars with passive restraints. They had only to acquire a production capacity, which they didn't have to utilize if consumer demand was insufficient. She argues further that if the auto companies were sincere about "generating consumer demand" for air bags, they could have done so on their own any time from 1970 to 1976.[86] Ralph Nader was perhaps the most critical, calling the Coleman plan "horrendous and irresponsible," "terrible and contradictory."[87]

Some pro–air bag congressmen questioned the need for a large-scale demonstration. Congressman John Moss said: "This is a meaningless, equivocal decision. Because it proposes no new safety standards, it is essentially a non-decision."[88] Representative John M. Murphy (D–N.Y.), who introduced a bill in January 1977 to require passive restraints, insisted that enough testing had already been done and that it was time to install the devices in all new cars.[89]

Some officials in the auto industry did not like the prospect of installing air bags but nonetheless expressed a grudging respect for Coleman as a man and for his plan. Roger Smith of GM in particular was not eager to renew the air bag program at GM but saw that the alternative was an across-the-board mandate. He thus supported the demonstration as a method to obtain a valid statistical evaluation of the technology's effectiveness.[90] Herb Misch of Ford, who participated in the negotiations and wanted to see the air bag in cars, was more enthusiastic. Misch had become a vocal supporter of the demonstration idea and was extremely disappointed when it was later overturned.[91]

GM's David Martin believes that mandatory installation of air bags on all cars in the mid-1970s would have been bad public policy. According to Martin, GM's Air Cushion Restraint System (ACRS) deployed too frequently at low speeds when it was not needed and was not very effective in preventing moderate injuries.

He saw the possibility of many injuries caused by air bag inflation in minor crashes. If these results had occurred on a fleet-wide basis, the "result would have been a storm of protest." In light of these considerations, Martin saw Coleman's demonstration plan as "a sensible idea."[92]

Among engineers and scientists at NHTSA, support was widespread—yet quiet—for Coleman's plan. One engineer at NHTSA explained: "There was a consensus in the technical community that Coleman's plan was a brilliant idea. It would resolve any lingering technical questions while building confidence in the technology. Once on the road in large numbers, the air bag would sell itself."[93] Gregory, who had resigned and was overruled by Coleman, was initially critical of Coleman but later spoke with admiration of his thinking: "I just didn't have the imagination of Coleman to extract such a commitment from the companies."[94]

Conclusion

The 1974–1977 period saw GM joining the rest of the auto industry in a determined effort to kill the passive-restraint regulation. Once Cole retired from GM, the change in corporate policy was swift and complete. The political debate then evolved into whether such an ambitious regulation in depressed times was economically reasonable.

The key weakness in the industry's strategy was that they offered no constructive safety program as an alternative to passive-restraint regulation. Political rhetoric from Detroit in favor of belt use legislation was not accompanied by a determined campaign by industry leadership to make things happen. Hence, the industry's strategy appeared too negative and self-serving.

An assessment of Gregory's administrative performance is difficult because the constraints he faced were enormous (e.g., depression in Detroit and antiregulation public sentiments) and the parties were sharply polarized. Although Gregory was clearly sympathetic with the case for passive restraints, he did not seem to have a coherent strategy for bringing the technology to the marketplace. A case can be made that Gregory should have been more receptive to Eads's demonstration proposal. The Eads plan could have been sold to Detroit as a joint NHTSA–White House plan, although there certainly would have been opposition from pro–regulation forces. If Gregory had been successful in completing a demonstration agreement in 1975, overturning it would have proved difficult for the new Democratic administration in 1977.

Coleman's administrative strategy was innovative. Rather than use regulation to compel technological progress, he used regulatory authority as a threat to induce voluntary action by auto companies. It was an especially appropriate strategy for an air bag advocate in a pro-business administration. Claybrook, for instance, believes that Coleman did as much as he could for air bags as part of the Ford administration and was "sincere in his support for the air bag."[95] According to Charles Livingston, who worked at NHTSA for fifteen years, "the peacemaker role was the right strategy for Coleman to take because regulation and highway safety had fallen from the political priorities of the nation."[96] The political wisdom of the Coleman agreements, says Brian O'Neill of the Insurance Institute for Highway Safety, was that "if Detroit failed to deliver their end of the bargain, Coleman or his successor could have publicly embarrassed the industry and then nailed their ass to the wall with regulation."[97]

If viewed as one decision, the Coleman compromise seemed weak and insufficient to ambitious air bag advocates. If viewed as the start of a dynamic, long-run challenge for both Detroit and NHTSA, the Coleman compromise looked promising. No one will ever know whether the compromise would have worked because it was never given a chance. In his confirmation hearings before the Senate, President Carter's nominee for secretary of transportation, Congressman Brock Adams, promised to "review most carefully" Secretary Coleman's handling of the passive-restraint issue.[98]

Endnotes

1. Interview with Joan Claybrook, July 24, 1987; interview with Brian O'Neill, July 24, 1987; interview with Don McGuigan, September 3, 1987.
2. "A Rehash of Air Bag Arguments," *Automotive Industries* (June 15, 1975), pp. 13–14; "Passive Restraint Meeting Held," *Status Report: Highway Loss Reduction* (June 18, 1975), pp. 1–15.
3. "Air-Cushion Restraints Called Unproven, Unpopular and Too Expensive," *Machine Design* (June 26, 1975), p. 10.
4. "Air Bag Suits: A Large Fear," *Automotive Industries* (November 15, 1974), p. 12.
5. Interview of Eugene Ambroso, September 3, 1987.
6. David E. Martin, "Automatic Restraints—A Ten Year Learning Curve," in *The Human Collision*, International Symposium on Occupant Restraint, Toronto, Canada, June 1981, p. 155.
7. Louis C. Lundstrom, "Integrating Vehicle Safety, Cost and Consumer Attitudes," *The Fourth International Congress on Automotive Safety*, U.S. Dept. of Transportation, NHTSA, Washington, D.C., July 1975, pp. 419–429; Martin, "Automatic Restraints—A Ten Year Learning Curve," p. 155.
8. *A Restrospective Analysis of the General Motors Air Cushion Restraint Sys-*

tem Marketing Effort, 1974 to 1976, Final Report of Booz, Allen and Hamilton, Inc. to NHTSA, July 1983, pp. v–vi.

9. Albert R. Karr, "Saga of the Air Bag, or the Slow Deflation of a Car Safety Idea," *Wall Street Journal* (November 11, 1976), pp. 1, 14.

10. Ibid.

11. Ibid.

12. Interview of David Martin, September 3, 1987; "Air-Cushion Restraints Called Unproven, Unpopular and Too Expensive," p. 10.

13. "Air Bag Suits: A Large Fear," p. 12.

14. Interview of David Martin, September 3, 1987.

15. *Motor Vehicle Facts and Figures 1984*, U.S. Motor Vehicle Manufacturers Association, Detroit, Michigan, 1985, pp. 16, 62.

16. Ibid., p. 42.

17. "A Rehash of Air Bag Arguments," pp. 13–14.

18. "Another Deflation for the Air Bag Fight," *Business Week* (April 28, 1975), p. 24.

19. "Congressional Leaders Urge NHTSA to Resist Moratorium," *Status Report: Highway Loss Reduction* (November 20, 1974), pp. 2–3.

20. "NHTSA Again Finds Air Bag Better than Belts," *Status Report: Highway Loss Reduction*, December 26, 1974, p. 5; "Analysis of Effects of Proposed Changes to Passenger Car Requirements of FMVSS 208," NHTSA, DOT, Washington, D.C., December 1974, DOT-HS-801-328.

21. Interview of James Gregory, July 14, 1987.

22. "Automotive Occupant Protective Safety Air Cushion Expenditure/Benefit Study for the Allstate Insurance Company," John DeLorean Corporation, August 1975; Charles Y. Warner, Michael R. Withers, Richard Peterson, "Societal Priorities in Occupant Crash Protection," in *Fourth International Congress on Automative Safety*, pp. 907–960.

23. Testimony of Donald Schaffer, vice president, Allstate Insurance Company, in *Installation of Passive Restraints in Automobiles*, Subcommittee on Consumer Protection and Finance, U.S. House of Representatives, 95th Congress, First Session, September 1977, p. 124.

24. Interview of James Gregory, July 14, 1987.

25. Ibid.

26. Ibid.

27. Executive Order No. 11821, November 24, 1974; Office of Management and Budget Circular A-107, January 28, 1975.

28. "DOT Proceeds with Air Brake Rule," *Status Report: Highway Loss Reduction* (January 21, 1975), pp. 6–9.

29. Interview with George Eads, September 3, 1987; interview with Thomas Hopkins, September 18, 1987; Lawrence Goldmuntz and Howard Gates, "Review and Critique of NHTSA's Passive Restraint System Cost-Benefit Analysis," Economics and Science Planning Inc., Washington, D.C., January 22, 1975.

30. Memorandum from George Eads, COWPS, to James Gregory, NHTSA administrator, February 7, 1975.

31. Testimony of George Eads, assistant director of government operations and research, COWPS, before the Consumer Subcommittee, Senate Commerce

Committee, March 20, 1975 (CWPS-32); testimony of George Eads, COWPS, Public Meeting on Occupant Crash Protection, NHTSA, May 23, 1975 (CWPS-49).

32. "Massive Air Bag Test Plan Suggested," *Status Report: Highway Loss Reduction* (March 31, 1975), pp. 7–8.

33. Ibid.

34. Interview of George Eads, September 3, 1987; interview of Thomas Hopkins, September 18, 1987.

35. Interview of James Gregory, July 14, 1987.

36. Ibid.

37. "Air Bag Test Plan 'Interesting' But . . .," *Status Report: Highway Loss Reduction* (May 12, 1975), pp. 6–7.

38. Ibid., p. 7.

39. "President Questions Air Bag Cost-Effectiveness," Ibid., p. 6.

40. Interview of Thomas Hopkins, September 18, 1987.

41. Interview of Howard Dugoff, August 4, 1987.

42. *The Secretary's Decision Concerning Motor Vehicle Occupant Crash Protection*, U.S. Department of Transportation, Washington, D.C., December 6, 1976, p. 21, note 20.

43. Interview of Robert Carter, August 17, 1987.

44. Interview of Howard Dugoff, August 4, 1987.

45. "Senator Asks Immediate Action on Passive Restraints," *Status Report: Highway Loss Reduction* (November 5, 1975), pp. 4–5.

46. Ibid.

47. "Moss Demands Timetable on NHTSA Rulemaking," *Status Report: Highway Loss Reduction* (February 3, 1976), pp. 2–3.

48. Interview of Lowell Dodge, August 13, 1987.

49. Interview of James Gregory, July 14, 1987.

50. Letter of Administrator James Gregory to Congressman John Moss, February 18, 1976.

51. Letter of Ralph Nader to Secretary William T. Coleman, December 23, 1975.

52. Letter of Secretary William T. Coleman to Ralph Nader, April 20, 1976.

53. Interview of James Gregory, July 14, 1987; "Gregory Resigns from NHTSA," *Status Report: Highway Loss Reduction* (March 3, 1976), pp. 4–5.

54. Testimony of James Gregory, *Regulatory Reform—Volume IV*, Hearings before Committee on Interstate and Foreign Commerce, U.S. House of Representatives, 94th Congress, 2nd Session, February 27, 1976, p. 431.

55. Interview of Don McGuigan, September 3, 1987.

56. Interview of Howard Dugoff, August 4, 1987.

57. Remarks of Secretary William T. Coleman, Economic Club of Detroit, May 24, 1976.

58. Interview with Brian O'Neill, July 24, 1987.

59. Ibid.; interview with Roy Haeusler, September 16, 1987.

60. Letter of Don Schaffer to Secretary William T. Coleman, May 25, 1976.

61. Press Release, Department of Transportation News, Office of the Secretary, June 9, 1976; *Federal Register*, 41:24070 (1976).

62. Morton Mintz, "Decision Delayed on Auto Safety Aids," *Washington Post* (June 10, 1976), p. 1.

63. Statement of Congressman John E. Moss, Chairman, Subcommittee on Oversight and Investigations of the Committee on Interstate and Foreign Commerce, U.S. House of Representatives, in *Regulatory Reform—Volume IV*, p. 525.

64. Morton Mintz, p. 1.

65. Ibid.

66. Letter of Secretary William T. Coleman to Ralph Nader, June 10, 1976.

67. Interview of Brian O'Neill, July 24, 1987.

68. Testimony of Ralph Nader, Public Hearing on Occupant Crash Protection, U.S. Dept. of Transportation, Washington, D.C., August 3, 1976.

69. Interview of Don McGuigan, September 3, 1987.

70. Interview of Howard Dugoff, August 4, 1987.

71. Ibid.

72. COWPS Testimony Before the Secretary of Transportation on "Air Bag" Proposal, August 3, 1976 (CWPS-174).

73. *The Secretary's Decision Concerning Motor Vehicle Occupant Crash Protection*, pp. 10, 52–57.

74. Ibid., pp. 52–57.

75. Ibid., pp. 6–7, 11–12.

76. Letter from the U.S. Attorney General's Office to the U.S. Department of Transportation, January 14, 1977.

77. Interview of Don McGuigan, September 3, 1987.

78. Susan J. Tolchin, "Air Bags and Regulatory Delay," *Issues in Science and Technology* 1:75 (Fall), 1984.

79. Ibid.

80. "How Coleman Sold Detroit on Airbags," *Business Week* (January 31, 1977), p. 36.

81. Interview of Don McGuigan, September 3, 1987.

82. *Federal Register*, 42:5071 (1977).

83. "How Coleman Sold Detroit on Airbags," p. 36.

84. Ibid.

85. Interview of Michael Finkelstein, July 23, 1987.

86. Interview of Joan Claybrook, July 24, 1987.

87. "Displeased with the Decision," *Status Report: Highway Loss Reduction* (December 13, 1976), p. 3.

88. Ibid., p. 1.

89. "Coleman Agreements Fall Short of Goal," *Status Report: Highway Loss Reduction* (February 3, 1977), pp. 1+.

90. "How Coleman Sold Detroit on Airbags," p. 36.

91. Interview of Helen Petrauskas, September 3, 1987.

92. Interview of David Martin, September 3, 1987.

93. Interview of Ralph Hitchcock, July 23, 1987.

94. Interview of James Gregory, July 14, 1987.

95. Interview of Joan Claybrook, July 24, 1987.

96. Interview of Charles Livingston, July 24, 1987.

97. Interview of Brian O'Neill, July 24, 1987.

98. "Adams Will 'Review' Plan," *Status Report: Highway Loss Reduction* (February 3, 1977), p. 5.

Chapter 6

CONGRESSIONAL STALEMATE

The U.S. Congress is one of the key power centers in modern American pluralism. The Senate and the House of Representatives share the power to make laws with the President. Yet Congress is itself a highly pluralistic institution, each body comprising a bewildering array of subcommittees and committees that share legislative authority. Although bills can on occasion be introduced and passed on the floor of both bodies without committee or subcommittee deliberation, the system is designed to make that outcome extremely rare. The result is a diffusion of power among dozens of subcommittee and committee chairmen.

In 1977 the Carter administration terminated the Coleman agreements and issued a mandatory passive-restraint standard that ultimately survived judicial review. The result was a four-year congressional battle between those who sought to weaken or repeal the rule and those who sought to protect or strengthen it. The emergence of this issue in Congress had actually begun with the interlock fiasco of 1974 but was inflamed when the Ford administration's policy was overturned by the incoming Democratic administration. The ensuing struggle in Congress during the late 1970s occurred at the same time the Iranian oil crisis was crippling the domestic auto industry and the public was growing weary of federal regulation.

The 1977–1980 period is frustrating because of the absence of a clear expression of public policy from the legislative branch of government. As we shall see, Congress refused to overturn the Carter administration's passive-restraint rule and then refused to save it, even when it appeared to be in grave jeopardy. The period ends with the failure of courageous efforts aimed at legislative compromise.

The New Democratic Team

President Jimmy Carter appointed and the Senate approved Congressman Brock Adams (D–Wash.) as secretary of transportation. Adams is a former prosecutor from the state of Washington and is described as a fair-minded, moderate politician. His style is to listen carefully to all of the arguments on a controversial issue and then seek a middle-of-the-road solution. According to Joan Claybrook, "Adams was a caring man who in his balanced way always sought to do the right thing."[1]

Those who criticize Adams say that he let stronger-willed people push him around. In stark contrast to Adams, President Carter's choice to run NHTSA, Joan Claybrook, was a powerful and provocative personality. In the previous chapters we saw glimpses of Claybrook's career: an aide to Congressman Mackay (D–Ga.) and later to Senator Walter Mondale (D–Minn.) during deliberations on the 1966 Safety Acts, administrative assistant to the first NHSB director, Dr. William Haddon Jr., and a key aide to Ralph Nader during the Nixon and Ford years. Prior to her political appointment at NHTSA, Claybrook's interests had expanded beyond auto safety into public interest organizing at a Nader organization called Congress Watch.

Claybrook was profoundly influenced by her two mentors, William Haddon and Ralph Nader. Her view of good safety policy reflected Haddon's public health model, while her tactical style reflected Nader's advocacy model. As a person, Claybrook is described as intelligent, friendly, emotionally volatile, articulate, and overwhelmed by values. According to one of her admirers at NHTSA: "Joan questioned the motives and the agenda of those who worked in Detroit. She saw government regulation as a vital force needed to balance the power held by manufacturers."[2] Another NHTSA official recalls that "Joan came to NHTSA with a mission and that mission was air bags."[3] From the day word of Claybrook's nomination was leaked in mid-February 1977, many auto industry officials had reservations and fears.[4]

Big Ambitions

The week after Secretary Coleman resigned, the staff to incoming Secretary Adams requested that NHTSA officials brief Adams on the passive-restraint demonstration agreements. The briefing was prepared by Frank Berndt, NHTSA's acting chief counsel, and

Howard Dugoff, a career NHTSA analyst (then associate administrator of policy) who had been a member of Coleman's ad hoc task
force and negotiation team. As Dugoff recalls his personal feelings,
"I went to the meeting very proud and enthusiastic about what we
had accomplished in the negotiations with the auto companies."[5]

After Dugoff made his briefing, it was clear that the audience—
Adams and his aides—were very skeptical of Coleman's reluctance
to issue a regulation. While they saw some value in a demonstration, they saw it only as part of a process that would certainly and
speedily consummate in an across-the-board regulatory requirement. On the way back to his office from the meeting, Dugoff was
told by John Snow (the departing NHTSA administrator) that "you
didn't make any friends with the new administration." Later, however, Adams recruited Dugoff to be NHTSA's deputy administrator
in order to bring some balance on this issue to the political leadership of NHTSA.[6]

In March 1977 Adams reopened the passive-restraint rule making in an effort to make a quicker and larger safety gain than was
expected to result from Coleman's compromise plan.[7] This action
vitiated the obligations of auto manufacturers to comply with the
terms of Coleman's demonstration agreement. The decision to re-
open the issue was made prior to Claybrook's Senate confirmation
although she heartily endorsed the move.

Adams and his staff made three arguments in support of the
decision to overturn the Coleman agreement. First, Adams was
concerned that "public resistance" to passive restraints—a key factor in Coleman's analysis—was not a statutorily permissible basis
for refusing to issue a standard under the 1966 Vehicle Safety Act.
Although the Coleman plan had not yet been challenged in court,
Adams preferred a policy that was consistent with the standard-
setting orientation of the 1966 act. Second, Adams disputed Coleman's prediction that mandatory passive restraints would trigger
substantial consumer resistance. Because the air bag was not a
"forced-action" system like the interlock, he saw no basis for fearing
another round of public and congressional backlash against the
agency. Finally, Adams feared that the demonstration program
would delay for five to eight years any DOT decision to order
passive restraints in all new cars. In light of the growing consumer
demand for small (and less crashworthy) cars, Adams was eager to
see motorists protected by passive restraints as soon as possible.[8]

Instead of implementing the Coleman plan, Adams called for
another round of public hearings on three policy options: continuation of voluntary use of manual lap/shoulder belts, state enactment
of mandatory seat belt wearing laws, and mandatory passive re-

straints. The new rule making caused all interested parties to reconsider their positions, including economists in the Council on Wage and Price Stability.

White House Economists Lay Low

When the Carter administration took office, economist Barry Bosworth of the Brookings Institution was appointed as the new director of the Council on Wage and Price Stability. He asked Tom Hopkins, an economist who had worked on COWPS since 1975, to be assistant director of government programs and regulations. In this capacity Hopkins was responsible for reviewing the economic basis of new rule makings.

Bosworth made clear to Hopkins that "some of COWPS's previous interventions in agency rulemakings were 'cheap shots' or were carried out in an unduly inflammatory fashion."[9] At the same time the Carter White House was under pressure from Congressman John Moss (D–Calif.) to limit COWPS's power to impose benefit-cost tests on consumer health and safety regulations.[10] Even among White House officials, there was "ambivalence" about whether COWPS's regulatory analysis functions should be continued.[11]

In this uncertain environment, Hopkins and his staff conducted a reanalysis of the occupant-restraint decision facing Secretary Adams. In comments submitted to NHTSA, Hopkins abandoned the council's previous support of a large-scale field test of air bags and enunciated a more proregulation position:[12]

> *In the Council's view, if DOT is satisfied that no serious technical problems exist with either air bags or passive belts, and this is a crucial "if," the available economic analysis appears to support developing a passive-restraint standard.*

In support of this view, Hopkins pointed to new studies of restraint-system effectiveness and more field experience with GM's air bag system—although the real-world data were still far less than COWPS had recommended in 1975 and 1976. Hopkins was convinced that there was no purpose in reasserting a pro-demonstration position since Secretary Adams had already rejected that alternative.[13]

Parties Remain Polarized

The testimony at DOT's public hearing in April 1977 indicated that the issue was just as polarized as it had been in May 1975.

Automakers and seat belt manufacturers supported mandatory belt use laws while insurers and consumer advocates urged passive restraints. No innovative policy alternatives were proposed.

Somewhat surprising testimony was offered by the representative from Eaton Corporation, Vice President Marshall Wright. As an air bag supplier, one might have expected Wright to support a regulation. He instead urged Adams not to order mandatory passive restraints:[14]

> *Nothing is so calculated to kill the air bag as a mandate. As one of the principal potential manufacturers, we tell you that no one, no one, has the necessary experience to go from zero to ten million air bags in one year. . . . There will be an unacceptable number of malfunctions and public attention will focus on them. The public will lose confidence in them and become hostile to the mandate.*

Eaton officials may have preferred Coleman's demonstration program for another reason: The contracts with Coleman specified that GM and Ford would test Eaton's type of product, air bags, rather than passive belts. If a mandatory performance standard were issued, Eaton had no assurance that air bags would be the method of compliance chosen by manufacturers.

Retired GM President Ed Cole also appeared at the April 1977 hearing as an expression of his enthusiasm for the air bag technology. He was still convinced that air bags were a better overall safety device than active belt systems. In a letter to William Haddon in early 1977, Cole explained:[15]

> *I personally have been directly involved with the air cushion system to know that much more can be done beyond the production system built to date. I suspect many refinements can be made and a substantial amount of cost savings effected from the designs which have been available to the public from the limited production.*

Cole's position was considered to be a significant symbolic victory for insurers and consumer advocates.

The Adams Plan

Adams considered seriously only one alternative to mandatory passive restraints. DOT policy analysts Don Trilling and an influential DOT engineer, Jack Fearnsides, were urging Adams to consider mandatory belt use laws. Claybrook responded with a strong memorandum to Adams about the political infeasibility of belt use laws.

She urged Adams to go with mandatory passive restraints, and he ultimately concurred with her opinion.[16]

In July 1977 Adams and Claybrook issued a final order mandating passive restraints. The rule provided that all new cars sold in the United States must have sufficient "passive" crash protection to protect occupants in a 30 mph frontal crash. This "performance standard" did not specify how cars had to be designed except that the protection offered must be automatic. The deadlines in the rule were model year 1982 for large cars, 1983 for mid-sized cars, and 1984 for small cars.[17]

Former Secretary William Coleman later described the new rule as "a grandstand play" because the performance standard did nothing to guarantee that the air bag technology would be produced and marketed. Coleman was also critical of the 1982 deadline. Because the effective date was after the 1980 presidential election, nothing prevented a future DOT secretary from altering or rescinding the rule before it took effect.[18]

Ralph Nader also criticized the new rule, albeit for somewhat different reasons than former Secretary Coleman. He saw no legitimate basis for not making the rule take effect at the start of the 1981 model year. He also saw no rationale for a phase-in period that caused the smaller cars—which needed the crash protection most— to receive passive protection last. Nader charged publicly that Adams and Claybrook (his former aide) had capitulated to industry's demands to soften the rule. GM in particular had sought a four- to six-year phase-in. Each year of delay, Nader argued, meant 10,000 unnecessary deaths and 100,000 unnecessary disabling injuries. Adams and Claybrook, Nader charged, showed "undue concern for the convenience of the manufacturers over the risks to motorists." He predicted that Detroit would use the extra lead time to overturn the standard politically.[19]

In retrospect, Ford officials believe that the timetable in the Adams plan assured that automatic belts would be chosen for compliance over air bags.[20] In that acceptable passenger-side air bags were not yet known to be feasible, Ford could not responsibly gamble on overcoming this significant technological risk. Full-front air bags might be ready by the last year of the phase-in, but that year covered small cars—the class size that most troubled air bag engineers. The Adams plan, moreover, lacked a provision for introduction of driver-side air bags as an interim step toward full-front air bag protection.

Claybrook insists that Adams decided on the 1982 deadline at the urging of others, despite her advice to the contrary.[21] Others add that Adams pushed the deadline back a year to satisfy White House officials who didn't want the rule to take effect in Carter's reelection year.[22] In any event, both Adams and Claybrook defended publicly the extended timetable. The gradual phase-in was intended to permit manufacturers to absorb the impact of introducing passive-restraint systems without undue technological or economic risk at the same time that they undertook efforts to meet upcoming emission and fuel-economy requirements. The lead times were longest for small cars because air bag development was least advanced for smaller vehicles.[23]

Given the longer lead time, Adams and Claybrook also requested that automakers continue to participate in the voluntary agreement established by former Secretary Coleman.[24] This request drew support from Eaton Corporation and other air bag suppliers, who wanted some real-world testing prior to a mandate covering the whole market. Adams had always been convinced that some real-world testing of passive restraints prior to the fleet-wide mandate would be desirable.[25]

General Motors and Ford declined DOT's request to implement the terminated Coleman plan. In a letter to Secretary Adams, GM Chairman Thomas Murphy indicated that the Coleman plan called for air bags to be installed in mid-sized GM cars, whereas the Adams rule called for passive restraints to be installed initially in large cars. The company had already expended substantial resources (engineers, laboratory space, and testing facilities) on mid-sized cars and was faced with the necessity of shifting highest priority to large cars.[26] Ford made a similarly negative response. The logistical problems were more severe at Ford, where the Coleman agreement had called for testing of driver-side air bags in compact cars.[27]

Ford and GM indicated later that they did intend to undertake some real-world testing of passive restraints prior to the 1982 model year. GM announced plans to offer automatic belts as an option on three lines of 1979 models, and air bags as an option on all full-sized cars at the start of the 1981 model year.[28] Ford announced plans to offer passive belts as a customer option on at least one mid-sized car in model year 1980 and air bags as an option on at least one full-sized car in model year 1981. And due to particular concern about technical problems with air bag protection in small cars, Ford also announced plans to offer passive belts as an option on at least one subcompact car in model year 1981.[29] Thus, the loss of the Coleman plan did not at the time seem to be a major drawback.

When Adams and Claybrook issued the new rule, they were promptly sued by both Public Citizen (a Nader group) and the Pacific Legal Foundation (a conservative public interest group).

Public Citizen asked the D.C. Circuit Court of Appeals to order an acceleration of the compliance deadlines in the new rule. Pacific Legal Foundation, in contrast, petitioned the court to overturn the rule because of insufficient real-world experience with air bags, possible air bag "hazards," and DOT's alleged failure to consider public opposition to the rule. Although a unanimous three-judge panel of the D.C. Circuit Court of Appeals ultimately upheld the new rule as issued, a more imminent and serious threat to the rule arose on Capitol Hill.[30]

Threat of Congressional Veto

When Adams and Claybrook decided to mandate passive restraints, they knew that a strong challenge to their decision would occur in the Congress. The anti-interlock legislation of 1974 foreshadowed this showdown by providing that Congress could overturn any future occupant-restraint rule through so-called "concurrent disapproval resolutions" (often called "legislative vetoes"). In particular, Congress had provided that if legislative-veto resolutions in both the House and Senate were passed within sixty calendar days, the new passive-restraint regulation would not become law.

The "antiregulation" forces in Washington had been growing more vocal throughout the 1970s. Some politicians—spurred by the writings of conservative economists—saw regulation as a contributor to inflation, which citizens regarded as public enemy number one. Other politicians saw regulation as the coercive force of "Big Brother"—a trend toward intrusion into the privacy and personal freedom of American citizens. And there were (as always) the congressional allies of auto manufacturing interests who sought to protect their industrial constituents from another burdensome regulation.

The changing political culture was accompanied and fostered by new faces on Capitol Hill. Consumer-safety advocates had already lost some of their heroes to electoral defeat and resignation—Vance Hartke (D–Ind.), Paul Douglas (D–Ill.), and Margaret Chase Smith (R–Maine)—while those advocates who remained were perhaps a bit more cautious. NHTSA in particular had suffered resounding legislative defeats in 1974–1975 on the interlock issue and on policy toward motorcyle helmet use (where NHTSA was stripped of its authority to withhold federal funds from states that refused to enact motorcycle helmet laws). During congressional debate on these issues, it was apparent that the bipartisan consensus behind auto safety regulation that emerged in the 1960s was beginning to unravel. The new generation of legislators elected in the 1970s had not been participants in the consensus of the 1960s and seemed to be responding to somewhat different political winds.

Opponents of mandatory passive restraints in Congress began to plot strategy as soon as Secretary Adams overturned the Coleman agreements in March 1977. They were led by Bud Shuster (R–Pa.) and John Dingell (D–Mich.) in the House of Representatives. Shuster is an energetic conservative from the small towns and rural farming regions of central Pennsylvania. He publicized the air bag as a hazard to consumers because the chemical to be used for inflation (sodium azide) was a carcinogen in animal tests.[31] Dingell is a savvy politician from the Detroit area who is not bashful about protecting the interests of the auto industry. He saw air bags as a bad consumer investment.[32]

On the same day in July 1977 that Adams and Claybrook announced their new regulation, a coalition of 162 opponents led by Shuster introduced a resolution of disapproval in the House. In the Senate, where opposition was somewhat less organized, disapproval resolutions were introduced by a junior Republican senator from Michigan (Robert P. Griffin) and a conservative Republican from Oklahoma (Dewey F. Bartlett).[33]

Supporters of passive restraints formed the National Committee for Automobile Crash Protection to lobby on behalf of the Adams plan. The committee represented thirty-four organizations and individuals ranging from insurers, medical societies, consumer groups, labor unions, and Ralph Nader. No major safety group lobbied against the Adams plan.

Conspicuously absent from this coalition was the National Safety Council (NSC), one of the oldest and largest safety groups in the country. NSC tended to take a more conservative posture on passive restraints, perhaps because its board of directors included several auto executives and its own programs tended to be oriented toward changing driver behavior. NSC's decision not to oppose mandatory passive restraints was actually an unexpected boost for Adams and Claybrook. According to Charles Hurley of NSC, "the case for regulation could have been hurt if a crucial group such as NSC had emerged as an anti–air bag force."[34]

Adams and Claybrook approached the congressional battle with some considerable political assets. Time was on their side in that their rule would become law unless both bodies of Congress disapproved it by October 19, 1977. Both houses of Congress had large Democratic majorities, and the newly elected Democratic President, Jimmy Carter, had made a public endorsement of the passive-restraint rule. Finally, both Adams and Claybrook possessed prior Washington experience and were able to call upon numerous contacts and friendships developed since the 1960s.

The Senate seemed to be particularly safe considering that Senator Warren Magnuson (D–Wash.)—an architect of the 1966 Safety Act and air bag advocate and an influential senior member—would chair the key committee on this issue (Commerce, Science, and Transportation). Adams and Claybrook, however, were worried about the House because the chairman of the critical Committee on Interstate and Foreign Commerce, Harley O. Staggers (D–W.Va.), was considered likely to give air bag opponents a fair hearing.[35]

The Controversial Calspan Crash Tests

Just before congressional hearings on the Adams plan began, Calspan Corporation of Buffalo, N.Y., performed the first so-called "offset" head-on crash tests involving cars with air bags. Accordingly, cars were lined up so that the drivers (or passengers) in each car were directly opposite each other (rather than aligned headlight to headlight). The NHTSA-sponsored tests were designed to compare the effectiveness of air bags and lap/shoulder belts in four offset collisions of cars traveling 30 mph. Cadavers and dummies were placed in the front seats of test cars.

The crash forces measured on the belted dummies were less than NHTSA's injury limits. However, the forces measured on two cadavers and a dummy in the air bag cars far exceeded the injury limits. A NHTSA scientist was quoted as saying, "NHTSA judges these occupants 'killed.' "[36]

Accusations ensued that NHTSA officials suppressed these test results to avoid unfavorable publicity about air bags. *Automotive News,* a Detroit-based trade publication, reported that "reprisals" were "threatened" against one of the participating scientists who tried to "make the results public."[37] Congressman John Dingell was especially perturbed about NHTSA's written summary of the test results. In Dingell's words: "The NHTSA staff summary is obviously not complete and has been highly edited. This might indicate possible inaccuracies; it might indicate better explanation or it might indicate they are cooking their facts downtown."[38]

Career NHTSA officials insist that no attempt was made to suppress the Calspan test results.[39] NHTSA's summary of the test results emphasized that "additional cadaver tests are needed to complete the assessment of the two systems. However, it is apparent that both systems provide lifesaving protection under the test conditions."[40] NHTSA analysts added that injuries in the cars with air bags could have been lessened if knee bolsters or lap belts had accompanied the air bag protection.

Tactics in the House and Senate

Public hearings on Shuster's resolution were held in September by the House Subcommittee on Consumer Protection and Finance of the Committee on Interstate and Foreign Commerce. Although the hearings were ceremonial to a large extent, they did provide a forum for both sides to showcase the work of their consultants, pronounce endorsements from sympathetic interest groups, and make their technical and political arguments. The real issue was who had the votes and who could make the best use of House rules to advance their interests.

Subcommittee chairman Bob Eckhardt (D–Tex.) was sympathetic with the Adams decision but recognized that many House members did not share his view. He did not feel it would be proper to kill the Shuster resolution in subcommittee, especially given the 162 cosponsors. Instead the subcommittee sent the resolution to the full committee with a negative recommendation. The subcommittee's report accompanying the resolution spoke favorably of the lifesaving promise of air bags.[41]

Committee Chairman Harley D. Staggers (D–W.Va.) wanted to send the resolution to the House floor, and he apparently led Dingell and Shuster to believe that he would do so. Eckhardt was prepared to accept that outcome, but wanted the full committee to be on record supporting the subcommittee's favorable report on air bags. An Eckhardt motion supporting the passive-restraint order was approved by voice vote.[42]

When Staggers tried to open committee debate on whether to send the issue to the floor, Henry A. Waxman (D–Calif.)—a pro-regulation liberal and ally of Claybrook and Adams—raised a point of order against any further committee action because the full House was supposedly "in session" to deliberate about amendments to a labor policy bill. (Committees are not permitted to hold meetings while the House is debating amendments to legislation, unless the committee receives special permission from the House.[43])

Waxman feared that if the Shuster resolution went to the House floor, Shuster and Dingell might assemble the votes needed to pass the disapproval resolution. For several days Staggers tried to convene the full committee, but the Waxman forces were absent at quorum calls, thereby preventing Staggers from getting a quorum. When the committee finally reconvened on October 12 (just a week before the expiration date), the members voted 16 to 14 to table the resolution; as a result, the Shuster resolution never reached the House floor.[44] Needless to say, Shuster and Dingell were bitter.[45]

In the Senate, efforts to overturn the Adams rule were also unsuccessful. The Consumer Subcommittee of the Commerce, Science, and Technology Committee voted 5 to 0 on September 29 to recommend that the committee table disapproval resolutions submitted by Senators Griffin and Bartlett.[46]

Anti–air bag forces were stronger in the full committee. Chairman Warren G. Magnuson (D–Wash.) tried to block further consideration by the full committee through agenda control. On October 6 Griffin forced a vote on his resolution by threatening to block committee action on any other legislation. The committee then voted 9 to 7—largely on partisan lines—to report the resolution to the Senate floor with a negative recommendation. Several days later a motion from the floor by Magnuson to table the resolution passed the Senate by a vote of 65 to 31.[47]

Opposition to the Adams plan was vocal and significant, but it was ultimately not strong enough to cause either body to pass a disapproval resolution. When the October 19 deadline passed without congressional action, the Adams plan became law. Opponents of the passive-restraint mandate, especially those in the House who felt they were victims of procedural manipulation, vowed to continue their fight against air bags. Representative John Dingell (D–Mich.) emerged as their leader.

In both 1978 and 1979 the House passed Dingell-sponsored amendments to DOT appropriations bills that barred DOT from spending taxpayer funds to enforce the passive-restraint mandate.[48] Although the Senate acquiesced to Dingell's provisions in conference committee negotiations, continued DOT funding of air bag research was permitted. The roll-call votes in the House indicated growing opposition to the Adams decision. The amendments had no practical effect, however, since the DOT rule was not scheduled to take effect until model year 1982.

Child Restraint Use Laws

While the federal government was embroiled in controversy about passive restraints, Tennessee became the first jurisdiction in the world to require that young childhood passengers be restrained in protective kiddie seats. A Tennessee pediatrician named Dr. Robert Sanders "marshalled the support of his colleagues throughout the state" to support child restraint use legislation.[49] NHTSA Administrator Joan Claybrook provided active support to the legislative effort in Tennessee during 1977.

Final passage of legislation in Tennessee is itself quite a saga. A strong bill fashioned by pediatricians appeared to have a good chance of passage. During floor debate, however, an unexpected amendment was offered to allow an exemption for mothers who are holding their babies in their arms, a potentially lethal practice. Authors of the original bill, seeing how popular the amendment was, saw the political reality as a weakened bill or no bill. Accordingly, they endorsed the weakened bill on the theory they could revisit the issue in future sessions and plug the loophole.[50]

Their judgment proved to be correct; the Tennessee law was ultimately amended. It was followed several years later by legislation in Rhode Island and other states. Beginning in 1979, the American Academy of Pediatrics hired several organizations to assist its state chapters in campaigns to enact child restraint use legislation. Working with NHTSA, pediatricians were mobilized in numerous states to lobby and testify for legislation. Claybrook played a very supportive role in the early phase of this campaign.

Pressure on Claybrook

During the period after NHTSA's July 1977 passive-restraint order, several congressmen sought to encourage use of safety belts among adults through economic incentives and mandatory seat belt wearing laws. Such laws had already taken effect in twenty-three countries and territories around the world, including Puerto Rico.[51]

Because the passive-restraint rule affected only new cars, it would be ten years before a majority of cars would have passive restraints and over twenty years before all cars on the road were equipped with automatic belts or air bags. During this long transition period, mandatory seat belt laws could save lives and reduce injuries in cars without passive restraints. Further, belt use advocates were afraid that air bags—by creating a false sense of security—might erode belt use levels, even though air bags and safety belts provide complementary crash protection.

In January and June of 1978, the House Subcommittee on Investigations and Review of the Committee on Public Works held public hearings on safety belt use in Puerto Rico and Washington, D.C.[52] This is the same committee that had authorized incentive grants for state belt use laws in 1973–1974 as part of the Highway Safety Act of 1973. The hearings were thus a logical extension of the earlier initiatives of Congressman William H. Harsha (R–Ohio), ranking minority member of the Public Works Committee. Harsha's interests were shared by three key subcommittee members: Chairman Bo Ginn (D–Ga.), Henry Nowak (D–N.Y.), and James Cleveland (R–N.H.). In addition to belt use laws, these congressmen were interested in whether tort liability awards and insurance policies could be modified to encourage safety belt use.

Enthusiastic testimony was offered at the hearings by Charles Pulley, President of the American Seatbelt Council, and by representatives from General Motors, Chrysler, and Ford. They urged the committee to authorize NHTSA to expand educational activities about belt use and to reinstate federal highway grant incentives to states that enacted belt use laws. Pulley also urged insurance companies to offer and publicize benefits for belt users; courts were urged to reduce liability awards to injured motorists who were unbelted at the time of the crash.[53] One of the world's leading experts on safety belts, Dr. B. J. Campbell of North Carolina's Highway Safety Research Center, also testified at the congressional hearing. He emphasized that the federal government had never given safety belt use a "fair trial" and that belt use was still not given "the high priority that it deserves." In light of the inhospitable political environment for across-the-board belt use laws, Campbell recommended that NHTSA focus on mandatory restraint use for children and beginning drivers. He cited Tennessee as the first state in the country that had covered young children with a mandatory restraint use law.[54]

Experts from the Insurance Institute for Highway Safety (IIHS) were not optimistic about raising manual belt use levels. Dr. Leon Robertson summarized scientific research showing that certain TV ads were not effective at increasing observed belt use rates. Mr. Benjamin Kelley cited public opinion polls that found a majority of Americans opposed to mandatory belt use. When asked whether insurance companies could modify policies to encourage belt use, the IIHS experts responded that there would be serious legal obstacles and, in addition, it would be difficult to evaluate the compliance and effectiveness of such policies.[55]

Committee members listened carefully to the testimony of NHTSA Administrator Joan Claybrook. She cited information on the low rates of voluntary belt use in the United States (20 percent and declining) and the ineffectiveness of various mass media and educational programs. She insisted that her staff had examined carefully and systematically every method available to encourage increased belt use. On the subject of international experience with compulsory belt-use legislation, she acknowledged that belt use rates vary from 50 to 90 percent depending upon the stringency of enforcement.[56]

Claybrook believed that public attitudes against belt use in the United States were so ingrained that "efforts by the DOT to encourage belt-use laws are unlikely to succeed."[57] The recent congressional opposition to motorcycle helmet use laws was an indication of the hostile attitudes faced by Claybrook. Her only encouraging remarks concerned child restraint laws, where she said NHTSA was prepared to play an actively supportive role—as it had done in Tennessee.[58]

Congressman Cleveland questioned Claybrook on the extent of NHTSA's educational efforts on adult belt use. He commented that the $6 million program (1970–1976) cited by Claybrook was a very small proportion of the $1.5 billion spent by NHTSA on all safety activities since passage of the 1966 Safety Acts. He concluded that promotion of belt use "looks to me as if it is a peanut operation down at your shop."[59] Cleveland was particularly disappointed that Claybrook's statement did not address the topic of financial incentives for the use of seat belts.[60]

Congressman John Fary (D–Ill.) went further and criticized Claybrook for devoting less than $300,000 in fiscal year 1979 to promoting belt use laws.[61] Claybrook responded that the House Appropriations Committee had "made it quite clear to us several years ago that they thought it inappropriate for us to spend federal funds and give them to the states to encourage mandatory belt usage laws." She added that NHTSA is prepared to offer "information and technical assistance, but not funds" to states that consider belt use legislation.[62]

Congressman Ginn responded that Claybrook and her predecessor in the job, James Gregory, had erroneously interpreted Congress's position on incentive grants for belt use laws. In a report dated July 22, 1975, the Senate Appropriations Committee stated:[63]

The Committee wishes to emphasize that no actions taken by the Congress in the past or present prevent NHTSA from developing incentive grant programs to encourage States to enact effective seatbelt usage legislation within the limits of available funds. The Committee is concerned that the NHTSA may have interpreted the limitation in conference last year of funds specifically earmarked for this purpose from the fiscal year 1975 Transportation Appropriations Bill as prohibition against any such program being instituted. Such is not the case.

Ginn added that NHTSA should have been pursuing seat belt laws for the last several years, despite opposition from certain members of the House Appropriations Committee.[64]

Claybrook responded that no funds had been authorized for incentive grants to states. She added that she could not support enactment of a mandatory federal belt use law because of the need for public participation and state and local enforcement. She recommended instead that Congress work to encourage belt use laws on an experimental basis in one or two states, with thorough monitoring and evaluation. She noted that Australia had enacted belt use legislation on a province-by-province basis.[65]

After the congressional hearings, Claybrook took only very limited steps to encourage belt use laws (e.g., she sent letters to each governor requesting a political assessment), an approach that served to dampen the little enthusiasm in the Congress on this issue.[66] Her lack of aggressive leadership on safety belt use was a major source of irritation among belt use advocates, auto industry officials, and officials from state safety programs. They saw her pessimistic attitudes as a self-fulfilling prophecy. One of Claybrook's aides at NHTSA who worked with state agencies acknowledged: "It is fair to say that Claybrook never made a dedicated effort to get mandatory belt-use laws."[67] Another aide offered the following explanation of her philosophy: "Joan didn't do much on mandatory belt use because her primary interests were in vehicle regulation. She was fond of saying that 'it is easier to get twenty auto companies to do something than to get 200 million Americans to do something.' "[68]

Unfulfilled Expectations

Adams and Claybrook expected that their new passive-restraint rule would cause large numbers of full-sized cars to be equipped with air bags. In particular, they expected all six-seat passenger cars to be equipped with air bags because automatic belts were not feasible for the front-center seating position. DOT projected in 1977 that the rule would ultimately cause 60 percent of new cars (mostly large ones) to be equipped with air bags and 40 percent with automatic belts (mostly small ones).[69]

Over a year after the rule was issued, it appeared that air bag development programs were moving forward rapidly. In a 1978 progress report on passive restraints, DOT reported that air bags would be offered ahead of schedule by Ford, GM, Volvo, and Chrysler while Volkswagen and Toyota were concentrating on passive belts.[70] A market survey by Peter D. Hart Research Associates found consumer interest in passive restraints and concluded that it "appears that there will be a sizable market for both air bags and automatic seat belts when consumers have a choice of passive restraint systems."[71] Meanwhile, Calspan Corporation was successfully crash testing an air bag restraint system for compact cars at impact speeds of 45 mph—a system designed specifically to prevent injury to out-of-position children.[72]

By late 1978, however, compliance plans had changed radically, and it became clear to Claybrook that their projection would prove to be wrong, at least in the short run. The Iranian oil crisis and spiraling gasoline prices were threatening the survival of Detroit's large-car market.[73] Automakers, now making plans to abandon the six-seat car, turned to the less expensive automatic belt as their primary compliance technology.

Allied Chemical Corporation, which had done extensive work for GM on air-cushion restraints, announced in December 1978 that it was "reluctantly" dropping out of the air bag business. The decision was based on the fact that GM and the other major manufacturers were concentrating on automatic belts in compliance plans.[74] Several months later Eaton Corporation made the same decision, despite spending $20 million in research on air bags over the previous thirteen years. Eaton Chairman E. Mandell deWindt confessed that "we have nothing but bruises to show for our efforts." The passive-restraint rule was no market guarantee, he said, because "you've got to wonder what happens when Brock Adams goes out." He said Eaton's management concluded that "we're better off taking our chances [with products] in the free market."[75]

As carmakers were turning away from air bags, Congressman John Burton (D–Calif.) held a press conference where he released three GM marketing studies indicating that many consumers prefer air bags to seat belts. The studies had been provided to Burton with a promise of confidentiality by GM Vice President David Potter.[76] A 1971 GM survey, which was made public in connection with a tort suit, showed that half of 630 potential new-car buyers said they preferred air bags to cumbersome seat belts after hearing a brief workshop on the costs and benefits of air bags. A 1978 GM survey revealed that consumers ranked air bags higher than seat belts in terms of "operation, comfort, and appearance." But most revealing were the results of a 1979 GM survey. About 70 percent of respondents selected air bags as the first choice among restraint systems, even though they were told air bags would add $360 to the price of a car.[77]

GM's head of technical liaison, William C. Chapman, responded to Burton's press conference by noting that "marketing studies are not considered to be infallible." He emphasized that "there is a lot of difference between saying I'm going to do something and writing a check to do it." Pointing to GM's unsuccessful air bag option on 1974–1976 models, Chapman said that real-world purchasing behavior of consumers does not indicate a substantial demand for air bags.[78]

Possible Dangers to Small Children

In October 1979 GM announced that it would delay introduction of air bags as an option due to concerns that small children may be injured when an air bag inflates on the passenger side.[79] In crash tests simulating an out-of-position child, GM engineers witnessed several pigs killed or severely injured by air bag deployment.[80] GM thus joined Ford and Chrysler, which had already decided against offering air bags on 1981 models.[81] Claybrook responded quickly by appointing a special team of NHTSA analysts to explore GM's concerns. She emphasized that GM's reservations are "very narrow" and seem to be based on "fragmentary and speculative" information.[82]

At Ford safety engineers were also very concerned about dangers to out-of-position occupants, especially the associated liability risks. Ford officials tried to persuade Claybrook and her aides to sponsor a public seminar on possible technical solutions to the problem of injury to out-of-position occupants.[83] She refused and instead directed her staff to produce a report on how important this problem might be. They found that the risk of injury to "out-of-position" children was very small.[84] Officials at Ford disagreed. According to one such official: "Claybrook never did enough work on this issue to resolve our concerns about passenger-side bags. Since we knew she would never approve driver-side bags alone, our company was driven to automatic belts as our primary compliance alternative."[85]

In December 1979 GM announced that its engineers had solved the possible hazards to out-of-position children. As a result the company renewed its plans to offer air bags, this time as an option on full-sized lines at the start of the 1982 model year. Ford officials were skeptical of the announcement because GM had not yet resolved the problem to their satisfaction.

Analysts at NHTSA concluded that GM's concern was a "real but minor" issue. They were suspicious that the concern was raised by GM as a rationalization for delay of the air bag offerings.[86] GM engineers deny these suspicions. They say that adjustments were made in the folding of air cushions and the rate of deployment to minimize the risk to children. The momentum behind child restraint use legislation also promised to reduce the number of out-of-position children.[87]

The Air Bag's Competitor

Volkswagen of America was the pioneer in design, development, and marketing of the automatic belt. Its system contained four elements: a shoulder belt; an emergency-locking, inertia-reel retractor that automatically adjusted belt length; an energy-absorbing knee bolster (instead of a lap belt); and an emergency-release button placed high near the door that is guarded by a starter interlock to encourage belt use. After extensive laboratory crash testing, engineers at Volkswagen became convinced that their shoulder belt/knee bolster combination provided occupant crash protection close to or better than that offered by the conventional lap/shoulder belt system.[88]

The VW passive belt was first introduced as an option on model year 1975 Rabbits. It was then made standard equipment in VW's top-of-the-line "L" model beginning with the 1976 model year. As this was one of VW's best-selling models, several hundred thousand Rabbits with passive belts were sold to consumers in the late 1970s.[89]

The new passive belt design was well received by consumers. Observational surveys sponsored by DOT consistently found usage rates around 80 percent for Rabbits with automatic belts.[90] Attitude surveys of Rabbit owners also indicated favorable customer reactions.[91]

Beginning in 1977 Richard Peet, president of Citizens for Highway Safety, had urged Volkswagen of America to advertise its automatic restraint system in television commercials. After persistent prodding, VW agreed to give it a whirl and asked Peet—a satisfied VW owner—to appear in the spot. An ad was prepared by Doyle Dane and Bernbach (VW's ad agency) and aired on cable and prime time television. This was the first automatic belt TV campaign ever undertaken.[92]

A complaint against VW was soon filed with the Federal Trade Commission on the grounds that the ad was deceptive. Preliminary inquiry by the FTC's regional office in New York caused VW to abandon the advertisement. Several years later FTC Commissioner George Douglas concluded that, upon investigation, the ad was not deceptive.[93] By then, however, VW had abandoned its safety campaign.

One of the essential features of any passive belt system is a mechanism that allows a motorist to escape from a locked belt system after a crash. This issue had been addressed initially in the early 1970s, but in July 1971 NHTSA proposed that passive belts be nondetachable to assure high usage rates. Concerns were raised that a nondetachable belt's emergency-release mechanism might be jammed in a crash, preventing a belt user from exiting the car.[94]

In early 1974 NHTSA responded to such concerns by requiring that all passive belts be detachable with a single buckle release.[95] To encourage high usage rates, the agency required further that such belts be guarded by starter-interlock systems.[96] This was the approach adopted by Volkswagen for the Rabbit. When Congress withdrew DOT's authority to require interlocks in 1974, NHTSA dropped the requirement that passive belts be guarded by interlocks (without banning the interlock). All passive belts still had to be detachable with a single buckle release.

In early 1978 GM, which was becoming increasingly interested in automatic belts, petitioned Claybrook to relax NHTSA's rules governing detachability.[97] GM wanted to try a nondetachable lap/shoulder belt system with a "spool" release for emergencies. The spool-release feature would be easily activated by a push-button mechanism. This feature would permit the belt to unwind from the retractor in an emergency, allowing sufficient slack for the door to be opened and the occupant to exit from the vehicle.

Claybrook approved the GM petition in late 1978.[98] Thinking that the GM design might achieve higher usage rates than detachable designs, she saw this action as a method to encourage more use-inducing belt systems. She also saw passive belts as a method to promote acceptance of air bags (since no buyer would be forced to purchase air bags if automatic belts were also available).[99]

Meanwhile, consumers were not reacting enthusiastically to GM's early automatic belt options. An interlock-guarded, detachable, automatic shoulder belt—plus a manual lap belt—was offered as a $50 option on the 1978 and 1979 Chevette. Only 11,000 units were sold.[100]

For model year 1980, GM introduced a detachable but nonseparable lap/shoulder belt system instead of the "spool-release" feature approved by Claybrook. The company offered the new system as a no-cost option for most of the model year and provided a $25 bonus to dealers for each car sold with the new belt design. Although about $1.5 million was spent on dealer showroom displays and other promotional efforts, a survey by the Insurance

Institute for Highway Safety found that few Chevettes with the new belt design were available in dealer showrooms.[101] Only 3 percent of the 415,000 Chevettes sold in model year 1980 were equipped with automatic belts. Attitude surveys found that a majority of users found the Chevette system to be cumbersome and uncomfortable.[102]

Paralysis at NHTSA

As the compliance decisions of the auto companies evolved in 1979 and 1980, it became clear to career NHTSA officials that the regulation would not bring air bags to the marketplace. Automatic belts were still worth something in terms of safety, but NHTSA officials were beginning to worry about disconnection rates. Some auto companies were looking at passive belt designs that allowed for easy and permanent detachment. In fact, GM offered such a system as an option on 1981 Cadillacs.

According to career NHTSA officials, Claybrook was indecisive at this critical juncture of the standard's implementation.[103] She needed to take some further steps to get air bags in cars, but she did not do so.

One explanation offered for her indecisiveness was her inability to use NHTSA's resources in the face of market and technical uncertainties. GM and Ford were not necessarily eager to commit to an ambitious and risky air bag venture while Claybrook, a regulator they distrusted, was running NHTSA. Bob Carter of NHTSA claims that "Joan Claybrook never understood that you cannot regulate the industry effectively unless they trust NHTSA's leadership and are willing to engage in a dialogue."[104]

Another explanation offered is that Claybrook was "charmed and persuaded by Pete Estes and Herb Misch."[105] Estes was president of GM and a protege of Ed Cole; Misch was Ford's veteran vice president in charge of safety engineering. They would make personal promises to Claybrook on air bags, renege on those promises, and then make new promises. "Before Joan saw through these empty promises," says one NHTSA official, "it was too late to do anything."[106]

Some NHTSA engineers wanted to respond to "the passive belt fad" by writing such restrictive "entry and egress" rules on passive belts that manufacturers would be forced to go to air bags. Claybrook supported the analytical effort behind this tactic but never took it to rule making.[107] Others wanted to encourage air bags by forcing companies to make automatic belts nondetachable.

Claybrook rejected this tactic, fearing that companies would install the coercive belt designs and the public would backlash against NHTSA as they had in the case of the interlock.[108] And according to Claybrook, the new DOT secretary who succeeded Brock Adams in 1979, Neil Goldschmidt, did not view air bags as a high priority and this made it difficult for Claybrook to launch a credible passive-restraint initiative.[109]

Those who defend Claybrook insist that in reality she had few options because she was confronted with an increasingly skeptical Congress.[110] Indeed, Claybrook was almost saved from her predicament by a hostile initiative on Capitol Hill. The growing air bag opposition in Congress was beginning to make headway, and she was compelled by these external forces to address the air bag issue squarely. What happened, as we shall see, is that Claybrook almost got air bags through her enormous expenditure of political energy.

Enter David Stockman

Opponents of mandatory passive restraints in the House of Representatives sought in late 1979 to kill the Adams plan once and for all by amending the Motor Vehicle and Cost Savings Authorization Act of 1980. A proposed amendment to the authorizations bill by Congressman David Stockman (R–Mich.) disallowed NHTSA from enforcing any occupant-restraint standard that did not provide consumers a choice between manual and passive systems. Stockman argued that his amendment provided consumers greater "freedom of choice." Opponents disputed this claim, stating the amendment defined "consumers" to include dealers, who had a track record of apathy toward passive restraints.[111]

Although Stockman was a congressman from Michigan, his amendment was no favorite among auto manufacturers. According to one Ford official, "Stockman's plan was totally unworkable; it was a manufacturer's nightmare."[112] In the extreme, Stockman's plan might mean that manufacturers would have to design a manual and passive system for every line of car. As the same Ford official explained, "help like that from the Hill we surely didn't need."[113] Stockman's plan seemed to be motivated more by ideological and philosophical concerns than by concern for Detroit's welfare.

The authorization bill ultimately passed the House with the Stockman amendment attached. Since no such provision was included in the Senate bill, an agreement was expected to be reached in conference by representatives of the House and Senate.

Claybrook, though put on the defensive by Stockman, launched a political counteroffensive. On ABC's "Good Morning America," Claybrook attacked Stockman's "consumer-choice" amendment on the grounds that manufacturers would be permitted to make cars "the way they are manufactured today." She said the "consumer-choice" language used by Stockman was "deceptive" because his amendment would give the ultimate choice to manufacturers and dealers rather than the public itself. Stockman responded that the intent of the amendment was "to guarantee" that "the American consumer has the right to go into that showroom and decide which kind of safety system he's going to buy."[114]

In March 1980—before a conference agreement was negotiated—General Motors informed Claybrook that the company had changed plans and would not offer air bags on any medium- or small-sized cars in the 1982–1986 model years. Detachable automatic belts would be used instead. The company originally said that it might still offer air bags as an option on some full-sized 1982 cars, but even this plan was short-lived. In June 1980 GM announced that it had postponed plans to provide air bags on full-sized cars until the 1983 model year, thereby saving $20 million in tooling costs. Ford had already announced plans to defer their air bag offerings on Lincoln Continentals and Mark IVs from model year 1981 to 1982. These delays came on the heels of the market instability caused by the Iranian oil crisis.[115]

Enter John Warner

When the House-Senate conferees met in late July, an unexpected senator emerged as the champion of the air bag. John Warner (R–Va.) had served previously as secretary of the Navy and was known as an ally of the Defense Department. He had also become a key ally in the domestic auto industry's campaign to slow imports and curtail regulatory burdens. Warner was concerned about the serious decline in the overall financial position of the domestic auto industry and the sharp declines in employment and sales.[116]

Claybrook, who was actively involved in the legislative negotiations, describes Warner as "a very charming Southern gentleman—a lady's man in the better sense of that phrase."[117] Like everyone else, Claybrook was surprised to learn that Warner was willing to fight for air bags. Some observers see Warner as a protechnology force in the military who saw the air bag in the same light.[118] He also saw highway deaths and injuries as a human tragedy and was persuaded by Claybrook and others that automatic belts would not do the job. Warner also told his aides that he liked air bags because his wife, Elizabeth Taylor, refused to wear safety belts.[119] As a result, Warner feared Stockman's amendment because it would not stimulate the availability of air bags—a position that infuriated Stockman.[120]

In the first of two meetings, House-Senate conferees rejected the Stockman plan and gravitated toward a plan devised by Warner in consultation with industry lobbyists and air bag advocates. Warner's plan called for delay of the standard from 1982 to 1983, reversal of the compliance schedule so that small cars would be covered (1983) before large cars (1984), and inclusion of an air bag requirement covering at least one "line of car" for each large manufacturer (defined as GM, Ford, Datsun, Toyota, and Volkswagen).[121]

Precious time was lost in an extended squabble over how a "car line" would be defined. Claybrook pushed for a broad definition so that air bags would have to be offered on a large number of cars. Industry lobbyists urged a narrow definition.

The details of the air bag plan were finalized at a second meeting of the conferees. Air bags were required on at least one car line produced by large manufacturers in no less than three of the model years between 1982 and 1986.[122]

The plan was also designed in a creative fashion to enlist the support of domestic manufacturers. Chrysler, which originally had been included as a "large" manufacturer, was dropped from the air bag provision. "Car line" was defined narrowly—over the objections of Claybrook—and was applied to any specific family of cars within a make that was similar in construction (e.g., GM's Cadillac de Ville and Coupe de Ville). The phase-in was delayed for one year so that Ford and GM would not have to expend resources on the design of passive restraints for large cars that were scheduled to be phased out or downsized after one year. Reversing the deadlines for small and large cars by one year was designed to help domestic manufacturers (who sold mostly large cars) and to get automatic restraints into small cars—the fastest growing segment of the market and the size class with the most urgent safety needs.[123]

When asked by a reporter how he could reconcile his air bag plan with his advocacy of help for the auto industry, Warner replied:[124]

Well, we're in mourning for 52 hostages being held in Iran, yet about 52,000 Americans lose their lives on the highway every year. That's an astonishing record and I think the airbag can help reduce those fatalities. Nobody will be forced to use the airbag, but it will be available in some cars by 1983 under this agreement.

While the domestic auto companies (especially President Pete Estes of GM) expressed support for Warner's plan, their enthusi-

asm was not great.[125] Ford's Washington representatives reportedly told headquarters in Dearborn that passage of the Warner
plan was inevitable.[126] At the same time, Warner was worried
about getting support from the domestic industry and therefore
characterized the rule as a competitive plus for America:[127]

> *We are forcing the foreign companies to get involved in safety in our
> country. With the country moving to more small cars and highway
> fatalities likely to increase, it is now or never for the airbag. I think it
> is going to be a tremendous sales force, and the American companies
> are far ahead of most foreign producers on airbag technology. I just
> hope the industry will be passive itself when the issue is back before
> the House and Senate.*

Completion of the conference agreement was complicated by
two seemingly unrelated issues: NHTSA's car bumper rule and a
legislative-veto provision for all new NHTSA rules. At the insistence of Senate Majority Leader Robert Byrd (D–W.Va.), conferees rolled back NHTSA's bumper strength standard from 5.0 to 2.5
mph—at least until mid-1983. This provision favored steel interests
at a Huntington, West Virginia, plant of Houdaille Industries at the
expense of aluminum interests in the state of Washington.[128] The
legislative-veto provision was added to gain the support of Senator
Harrison Schmitt (R–N.Mex.), who saw legislative vetoes as a critical issue in the "regulatory reform" movement.[129]

Counting Votes

On the Senate floor, Senator Warner made a strong case for the
conference substitute. Senator Jesse Helms (R–N.C.) rose in defense of the House-passed Stockman amendment. When Helms
moved to table (and thereby kill) the conference substitute, he was
defeated by a vote of 30 to 56. The Senate then voted to adopt the
conference substitute by a vote of 48 to 38.[130] Warner was successful in marshaling a majority of both Republicans (18 of 35) and
Democrats (30 of 51) in support of his plan.

In the House of Representatives, John Dingell (D–Mich.) and
James Broyhill (R–N.C.) led a spirited fight against Warner's plan.
They argued vehemently against the air bag provision on the
grounds that it was a "design" rather than a "performance" standard.[131] The 1966 Vehicle Safety Act expressed a clear preference
for performance standards because manufacturers were viewed as
better situated than regulators to make design choices.

Dingell and Broyhill also gathered votes from some air bag supporters who were bitter about the Senate's insistence on a weak bumper rule. For example, Representative Bob Eckhardt (D–Tex.) urged the House to turn down the conference report because Senate conferees had held passive restraints "hostage" until the bumper protection was weakened. Eckhardt explained that "the reason I oppose this bill is because I do not believe we should be forced to barter away one [crashworthy bumpers] to get the other [air bags]."[132]

On October 1, Representative James Scheuer (D–N.Y.), an air bag supporter, made a motion to suspend the rules and adopt the conference substitute containing Warner's plan. The House voted 209 to 192 in favor of Scheuer's motion, but the winning margin did not satisfy the two-thirds required to suspend the rules.[133] Despite this discouraging defeat, Claybrook prepared strategy for another floor vote.

The issue was revisited two months later when Congress returned for the lame-duck session following the 1980 elections. With the Reagan administration about to take office, Claybrook worked hard to pass the Warner plan, despite the apathy of DOT Secretary Neil Goldschmidt. She enlisted the support of Speaker Tip O'Neill and the House Rules Committee in a last-ditch effort to get the conference substitute before the House for a vote.[134]

On December 4, the House considered a motion to open debate on the conference substitute, the necessary step toward passage. Since reconsideration of the conference report was to occur this time under regular procedures, only a majority vote was required. The motion failed by three votes, 186 to 189.[135] According to Michael Finkelstein, who worked with Claybrook at NHTSA, "Dingell simply beat us."[136] Claybrook recalls with sadness that some of her strongest allies happened to be absent when the House leadership finally allowed a vote on reconsideration of the conference report.[137] At this point, Claybrook felt all had been lost.

The next day Dingell and Broyhill offered a new bill that was virtually identical to the conference substitute except that it dropped the air bag and legislative-veto provisions. It was rejected by voice vote.[138]

A Last-Ditch Effort

On the day before the Senate was to go out of session, Senator Warner and the staff of the Senate Commerce Committee fashioned a new compromise bill. The hope was that some slight concessions to Dingell would garner enough extra votes to pass the bill in the House. As a face-saving measure for Dingell, a provision was added that required auto companies to add a line-item price of air bags on the window sticker of each new car with air bags. Dingell—who was under pressure from Byrd to allow the bill to pass—liked this window-sticker provision because it would make clear to consumers how costly air bags are. [139]

On the evening of December 16 the compromise bill was submitted to the Senate leadership with knowledge that its passage had been "greased." Both the auto industry and the Carter administration (Goldschmidt and Claybrook) were supportive. After Senate approval Dingell was prepared to allow it to pass in the House on a voice vote. All the lobbyists and administration officials had gone home. [140]

At 11:00 P.M. Cindy Douglas of the Senate Commerce Committee staff called Claybrook at home and told her she had better get up to the Hill "ASAP." Senator Howard Metzenbaum (D–Ohio) had "placed a hold" on the new bill. (Metzenbaum is a maverick liberal with close ties to organized labor and Nader.) Since the Senate was scheduled to go out of session the next morning at 5:00 A.M., Metzenbaum was prepared to kill the new bill by filibuster. Metzenbaum objected to the bill because he said it was wrong to weaken the bumper standard. [141]

Claybrook called Nader, and both of them came to the Hill. Warner urged Nader to try to persuade Metzenbaum to let the bill go through, but Nader surprised everyone by refusing to tell Metzenbaum it was a good bill.

At a late-night meeting with Senator Warner and Cindy Douglas, Nader explained that "you can never lower your bottom line." He objected on principle to both the window-sticker provision and the bumper provision (which had been in the plan since September). Warner was furious. Douglas told Nader that "the bill could save 9,000 lives" and if it is not passed, the incoming Reagan administration "is likely to rescind the entire standard." She explained further that if Nader opposes this pro-safety bill, he "will never be able to say credibly that Dingell has blood on his hands." Nader insisted that "the bill isn't worth it." If the new administration tried to rescind the rule, that "will serve to rekindle the consumer movement," Nader said. [142]

When President Ronald Reagan entered office in January 1981, the Adams plan was still in effect. All large cars produced after September 1, 1981, were to be equipped with automatic restraints. The legislative efforts of Senator Warner appeared to be wasted. Meanwhile, the Reagan administration's incoming director of OMB, David Stockman, was advocating repeal of the entire passive-restraint rule—an action that he estimated would save Detroit $100 to $160 million over the next three years. [143] Stockman told reporters that the air bag is "a dead issue." [144]

Conclusion

One of the central lessons of Claybrook's tenure at NHTSA is that a "technology-forcing" performance standard was not a very successful strategy for promoting installation of the air bag. By its very nature, the performance standard left technological discretion in the hands of the producers. For an administrator such as Claybrook who desperately wanted air bags, a different administrative strategy was required. Claybrook could have either retained the Coleman plan and acted as the government's enforcer of GM's and Ford's promises to market air bags, or taken the high-risk strategy of promulgating an air bag design standard (i.e., a rule that compelled carmakers to install air bags). Claybrook ultimately fought for the latter solution on the Hill against enormous odds and narrowly lost.

The more general lesson is that it is difficult for an NHTSA administrator to be influential if the entire industry is distrustful of him or her. NHTSA simply lacks the resources—budgetary, technical, legal, and political—to wage persistent battles with a unified Detroit. Even a highly energetic, proficient Washington insider like Claybrook cannot count on beating recalcitrant manufacturers on the Hill when appeals are made to legislators for relief from NHTSA rules. A more subtle administrative strategy is required, perhaps one that pits one large manufacturer against competitors. The Warner plan, with its anti-import and prodomestic character, was one type of strategy that was capable of enlisting some corporate support.

From the perspective of corporate policy, the discouraging experience of air bag suppliers offers some constructive warnings. The supplier who must count on the government to stimulate demand for his or her product is in a highly vulnerable position. Insofar as Eaton and Allied-Chemical spent time lobbying Washington instead of Detroit, they placed their substantial R&D investments in the air bag at considerable risk.

Perhaps the most significant lesson of the 1977–1980 period is that Congress, with its fragmentation and short attention span, is not a good power center to serve as the forum for auto safety regulation. Hurried, late-night negotiations between House and Senate conferees are not a sound process for deciding what technological innovations should be designed into new cars. An expert administrative agency, NHTSA in this case, is the best institutional forum for translating congressional goals into specific regulatory programs. Unless NHTSA's political leadership is exceptionally incompetent or untrustworthy, Congress should defer to NHTSA on specific regulatory matters.

Endnotes

1. Interview of Joan Claybrook, July 24, 1987.
2. Interview of Michael Finkelstein, July 23, 1987.
3. Interview of Barry Felrice, July 23, 1987.
4. "Give the Lady a Chance," *Automotive News* (February 21, 1977), reprinted in *Status Report: Highway Loss Reduction*, (February 21, 1977), p. 5.
5. Interview of Howard Dugoff, August 4, 1987.
6. Ibid.
7. *Federal Register*, 42:15935–15937 (1977).
8. Ibid.
9. Interview of Thomas Hopkins, September 18, 1987.
10. "Carter Warned About Wage and Price Control," *Status Report: Highway Loss Reduction* (March 29, 1977), pp. 8–9.
11. Interview of Thomas Hopkins, September 18, 1987.
12. Comments of the Council on Wage and Price Stability, Occupant Car Crash Protection FMVSS 208, May 31, 1977 (CWPS-244), p. 2.
13. Interview of Thomas Hopkins, September 18, 1987.
14. "Don't Order Mandatory Air Bag Use, Urges Eaton—Its Inventor," *Iron Age* (May 9, 1977), p. 12.
15. Letter from Edward N. Cole to William Haddon, Jr., January 4, 1977.
16. Interview of Joan Claybrook, September 24, 1987; interview of Howard Dugoff, August 4, 1987.
17. *Federal Register*, 42:34289–134299 (1977).
18. "Question Raised by GM Retraction," *Washington Post* (June 6, 1980), pp. F1–2.
19. Testimony of Ralph Nader in *Installation of Passive Restraints in Automobiles*, Hearings before Subcommittee on Consumer Protection and Finance, Committee on Interstate and Foreign Commerce, U.S. House of Representatives, 95th Congress, First Session (September 1977), pp. 252–272.
20. Interview of Don McGuigan, September 3, 1987; interview of Roger Maugh, July 8, 1987.
21. Interview of Joan Claybrook, July 24, 1987.
22. Interview of Carl Nash, July 23, 1987.
23. "Economic Impact Assessment," Amendment to FMVSS 208: Occupant Crash Protection, U.S. DOT, NHTSA, July 1977; *Federal Register*, 42:34295–34296 (1977).
24. *Federal Register*, 42:34295–34296; 1977; letter from Secretary Brock Adams to GM Chairman Thomas A. Murphy, June 30, 1977.

25. Interview of Howard Dugoff, August 4, 1987.
26. Letter from GM Chairman Thomas A. Murphy to Secretary Brock Adams, August 24, 1977.
27. Testimony of Herbert Misch, vice president of environmental and safety engineering, Ford Motor Company, in *Passive Restraint Rule*, Hearings Before the Subcommittee on Consumers, Committee on Commerce, Science and Transportation, U.S. Senate, 95th Congress, First Session, September 1977, p. 65.
28. "Broader Air Bag Program Announced by GM," *Iron Age* (September 5, 1977), p. 18.
29. Testimony of Herbert Misch, p. 67.
30. *Pacific Legal Foundation* vs. *Dept. of Transportation*, 593 F2d (D.C. Circuit 1978), p. 1338.
31. Letter from Congressman Bud Shuster to Congressman Wendell Ford, September 12, 1977.
32. Testimony of Congressman John Dingell, in *Installation of Passive Restraints in Automobiles*, pp. 179–197.
33. "Air Bags Mandated," *Congressional Quarterly Almanac* (1977, 1978), p. 532.
34. Interview of Charles Hurley, July 22, 1988.
35. Interview of Carl Nash, July 23, 1987.
36. Ed Janicki, "Air Bags: Mandated Market," *Automotive News* (February 29, 1988), p. E4.
37. Ibid.
38. Testimony of Congressman John Dingell, *Installation of Passive Restraints in Automobiles*, Hearings Before the Subcommittee on Consumer Protection and Finance, Committee on Interstate and Foreign Commerce, U.S. House of Representatives, 95th Congress, First Session, September 1987, p. 185.
39. Interview of Michael Finkelstein, July 23, 1987.
40. "Calspan Test Data: Summary of Recent Crash Tests of 1973 Chevrolets," NHTSA, September 1977, reprinted in *Installation of Passive Restraints in Automobiles*, pp. 34–38.
41. "Air Bags Mandated," p. 532.
42. Ibid.
43. Ibid.
44. Ibid.
45. Testimony of Congressman John Dingell, *Installation of Passive Restraints in Automobiles*, pp. 184–185.
46. "Air Bags Mandated," p. 532.
47. Ibid.
48. "CQ House Votes," *Congressional Quarterly Almanac* (1979), p. 126-H.
49. Speech by Joan Claybrook, "Stretching the Physician's Role: Influencing Health Policy by Giving Priority to Injury Prevention," Stanford University, January 24, 1986.
50. Interview of David Martin, September 3, 1987.
51. *Seat Belt Use Abroad*, American Seat Belt Council, Washington, D.C., June 1978.
52. *Safety Belt Use*, Hearings before Subcommittee on Investigation and Review, Committee on Public Works and Transportation, U.S. House of Representa-

    tives, 95th Congress, Second Session, January 4 and 5, 1978 and June 6, 7 and 8, 1978.

53. Testimony of Charles Pulley, ibid., pp. 199–242; testimony of John G. Manikas (Ford), Roland A. Quellette (GM) and Christopher Kennedy (Chrysler), ibid., pp. 336–346.

54. Testimony of B.J. Campbell, ibid., pp. 361–378.

55. Testimony of Leon Robertson and Benjamin Kelley, ibid., pp. 243–277.

56. Testimony of Joan Claybrook, pp. 136–151.

57. Ibid., p. 144.

58. Testimony of Joan Claybrook, *Safety Belt Use*, p. 144.

59. *Safety Belt Use*, pp. 148–149.

60. Ibid.

61. Ibid., p. 167.

62. Ibid.

63. Report of the Senate Appropriations Committee, July 22, 1975, excerpts reprinted in ibid., p. 377.

64. Ibid.

65. Testimony of Joan Claybrook, pp. 378–380.

66. "States Urged to Consider Mandatory Belt Use Laws," *Status Report: Highway Loss Reduction* (September 20, 1978), pp. 5–6; "Governors See Little Hope for Belt-Use Laws," *Status Report: Highway Loss Reduction* (December 14, 1978), pp. 1–3.

67. Interview of Charles Livingston, July 24, 1987.

68. Interview of Barry Felrice, July 23, 1987.

69. "Economic Impact Assessment," July 1977, pp. 2, 8.

70. "Surveys Show Consumers Prefer Air Bag Protection," *Status Report: Highway Loss Reduction* (September 20, 1978), pp. 1–4.

71. *Public Attitudes Toward Passive Restraint Systems*, U.S. DOT, NHTSA, Final Report by Peter D. Hart Research Associates, August 1978, DOT-HS-803-567, p. 7.

72. "Air Bag 'Tuned' to Small Car," *Machine Design* (February 9, 1978), p. 6.

73. *Motor Vehicle Facts and Figures 1984*, U.S. Motor Vehicle Manufacturers Association, Detroit, Michigan, 1985, p. 16.

74. Donald D. Holt, "Why Eaton Got Out of the Air Bag Business," *Fortune* (March 12, 1979), pp. 146–49.

75. Ibid.

76. Interview of Eugene Ambroso, September 3, 1987.

77. *Congressional Record*, December 12, 1979, p. E6094.

78. "GM Air Bag Study Said Suppressed," *Durham Morning Herald* (December 7, 1979), p. 11A.

79. "GM Will Delay Introduction of Air Bags on Some 1981 Cars Due to Safety Concerns," *Wall Street Journal* (October 2, 1979), p. 3.

80. Interview with David Martin, September 3, 1987.

81. Ibid.

82. "Joan Claybrook's Statement," *Status Report: Highway Loss Reduction* (October 9, 1979), pp. 3–5.

83. Interview with Roger Maugh, July 8, 1987.

84. Interview with Michael Finkelstein, July 23, 1987.

85. Interview with Roger Maugh, July 8, 1987.
86. Interview of Michael Finkelstein, July 23, 1987.
87. Interview of David Martin, September 3, 1987.
88. Testimony of Philip A. Hutchinson, Jr., vice president for Industry-Government Relations, Volkswagen of America, Public Hearing on Occupant Crash Protection, U.S. Department of Transportation, NHTSA, Washington, D.C., December 6, 1983, p. 4.
89. Ibid., pp. 4–5.
90. Opinion Research Corporation, *Automatic Safety Belt Systems: Owner Usage and Attitudes in GM Chevettes and VW Rabbits*, U.S. DOT, NHTSA, Washington, D.C., May 1980, DOT-HS-805-399.
91. Ibid.
92. Testimony of Richard C. Peet, president, Citizens for Highway Safety, Public Hearings on Occupant Crash Protection, U.S. DOT, NHTSA, Washington, D.C., December 7, 1983, pp. 7–8.
93. *Automotive News* (July 19, 1983), p. 1.
94. *Federal Register*, 36:12,858 (1971).
95. *Federal Register*, 39:3,834 (1974).
96. *Federal Register*, 39:14,593 (1974).
97. *Federal Register*, 43:21913 (1978).
98. *Federal Register*, 43:52493 (1978).
99. Interview of Michael Finkelstein, July 23, 1987.
100. David E. Martin, "Automatic Restraints—A Ten Year Learning Curve," in *The Human Collision*, International Symposium on Occupant Restraint, Toronto, Canada, June 1981, pp. 160–161.
101. "Survey Finds Scarcity of Automatic-Belt Chevettes," *Status Report: Highway Loss Reduction* (March 26, 1980), pp. 1–3.
102. Opinion Research Corporation, 1980.
103. Interview of Ralph Hitchcock, July 23, 1987; interview of Michael Finkelstein, July 23, 1987.
104. Interview of Robert Carter, August 17, 1987.
105. Interview of Michael Finkelstein, July 23, 1987.
106. Ibid.
107. Ibid.
108. Ibid.
109. Interview of Joan Claybrook, July 24, 1987.
110. Interview of Carl Nash, July 23, 1987; interview of Michael Finkelstein, July 23, 1987.
111. "The Big Air Bag Debate," *Washington Star* (February 4, 1980), p. A6.
112. Interview of Roger Maugh, July 8, 1987.
113. Ibid.
114. "Stockman Insists He Doesn't Want to Outlaw Air Bags," *Status Report: Highway Loss Reduction* (June 10, 1980), pp. 1, 8.
115. "GM Again Reverses on Air Bag Commitment," *Status Report: Highway Loss Reduction* (June 10, 1980), pp. 1, 10.
116. "Airbag Bill Revives with an Unexpected Champion on the Hill," *Washington Post* (August 7, 1980), p. A15.
117. Interview of Joan Claybrook, July 24, 1987.

118. Interview of Michael Finkelstein, July 23, 1987.
119. Interview of Cindy Douglas, August 12, 1987.
120. Ibid.
121. "Congress Nears Decision to Require Air Bags on Some Cars in Two Years," *Wall Street Journal* (July 31, 1980), p. 19.
122. "Conferees Adopt Air Bag Plan," *Status Report: Highway Loss Reduction* (August 6, 1980), pp. 1, 3–6.
123. Statement of Senator John Warner, *Congressional Record* (September 25, 1980), pp. S13501–503.
124. "Airbag Bill Revives with an Unexpected Champion on the Hill," p. A15.
125. Letter of GM President Pete Estes to Secretary Neil Goldschmidt, July 24, 1980.
126. Interview of Don McGuigan, September 3, 1987.
127. "Airbag Bill Revives with an Unexpected Champion on the Hill," p. A15.
128. Statement of Senator Robert Byrd, *Congressional Record* (September 25, 1980), p. S13501.
129. Statement of Senator Harrison Schmitt, ibid.
130. "CQ Senate Votes," *Congressional Quarterly Almanac, 1980* (1981), p. 59-S.
131. "Air Bag Bill Fails," *Congressional Quarterly Almanac, 1980* (1981), p. 265.
132. "Air Bag and Bumper Decisions Stalled," *Status Report: Highway Loss Reduction* (October 9, 1980), p. 5.
133. "CQ House Votes," *Congressional Quarterly Almanac, 1980* (1981), p. 156-H.
134. Interview of Joan Claybrook, July 24, 1987.
135. "CQ House Votes," *Congressional Quarterly Almanac, 1980* (1981), p. 170-H.
136. Interview of Michael Finkelstein, July 23, 1987.
137. Interview of Joan Claybrook, July 24, 1987.
138. "Air Bag Bill Fails," *Congressional Quarterly Almanac, 1980* (1981), pp. 265–266.
139. Interview of Cindy Douglas, August 12, 1987; statement of Senator John Warner, *Congressional Record* (December 16, 1980), pp. 34335–34336.
140. Ibid.
141. Ibid.
142. Ibid.
143. Susan J. Tolchin, "Air Bags and Regulatory Delay," *Issues in Science and Technology* 1:78 (Fall 1984).
144. "Reagan Victory May Accelerate Auto Recovery," *Ward's Auto World* (December 1980), p. 31.

Chapter 7

"REGULATORY RELIEF" AND THE SUPREME COURT

One of the central functions of the federal judiciary in America's constitutional government is to provide an independent check on abuses by the Congress and the executive branch. Judges are appointed to lifelong terms in order to insulate them from political pressure and majoritarian excesses. By enforcing the laws of Congress and our Constitution, the federal judiciary safeguards the fundamental rights and political benefits secured by individuals through pluralistic competition.

There is perhaps no better illustration of the federal judiciary's capacity to influence public policy than the history of the passive-restraint rule making. In 1972 the Sixth Circuit Court of Appeals made a searching review of NHTSA's passive-restraint rule and, while upholding the NHTSA's authority to regulate, suspended it due to technical inadequacy. In 1978 the D.C. Circuit Court of Appeals reviewed the same rule accompanied by better technical justification and upheld it. The correctness of these decisions as a matter of law is not seriously questioned in the legal literature.

When the Reagan administration sought to rescind the passive-restraint rule in 1981, there were good reasons to believe that the federal courts would not play a disruptive role. The notion of courts blocking new regulations was much more acceptable than the notion of courts blocking deregulation. There were few precedents for judges blocking the ambitions of an agency that was determined not to regulate. At the same time, the Burger Court's political conservativism seemed to be compatible with Ronald Reagan's 1980 campaign promise of "regulatory relief." The public mood was against regulation, Congress seemed apathetic, and a popular President was simply carrying out the platform that had reaped him a landslide victory at the polls.

Virtually everyone was surprised when the "conservative" Burger Court stopped the Reagan administration dead in its tracks in a unanimous decision. The deregulators were reminded that the federal judiciary is an independent source of power in the American political system. By keeping the passive-restraint issue on the nation's political agenda, the Supreme Court provided a new opportunity for America to achieve safety progress.

Depression in Detroit

When Ronald Reagan entered the White House in January 1981, the domestic automobile industry was in the midst of its deepest slump since World War II. Retail sales of domestic automobiles fell from 9.3 million units in 1978 to 8.3 million in 1979 to 6.6 million in 1980.[1] General Motors, Ford, and Chrysler reported record losses of $4.0 billion in 1980.[2] The unemployment rate in Michigan rose from 6.9 percent in 1978, to 7.8 percent in 1979, to 12.6 percent in 1980.[3] The automotive sector of the economy was much more depressed than the overall economy.

The Arab oil embargo of 1973–1974 and the Iranian oil crisis of 1979 had disrupted consumer confidence in the economy. The adverse effects of these oil shocks included 20 percent interest rates, a double-digit rate of inflation, and a sharp decline in real consumer income.[4] These effects compounded Detroit's problems by depressing total car sales and shifting the market away from (fuel-inefficient) domestic models.

As domestic manufacturers scrambled again to modify production capabilities for fuel-efficient cars, they found that they could not accomplish the profit margin on small cars that they had become accustomed to with large cars. Yet consumers wanted more small cars. The proportion of retail car sales accounted for by subcompacts and compacts had risen from 23.5 percent in 1972 to 36.4 percent in 1980.[5] The domestic manufacturers had grown accustomed to a relatively captive large-car market in North America and ill-equipped to respond to the rapid shift in demand for small cars. The market winners were importers (largely Japanese) whose share of the new-car market climbed from 14.7 percent in 1972, to 18.5 percent in 1978, to 26.7 percent in 1980.[6]

In the midst of this depression, Roger Smith became chairman of the board of General Motors Corporation. The company was experiencing its first annual loss in sixty years. From the outset,

Smith made clear that his vision of GM's recovery included fewer burdensome government regulations, such as the passive-restraint standard.

Calls for "Regulatory Relief"

The economic hardship in Detroit and throughout America caused politicians in Washington to scramble for remedial measures. The Reagan administration, which had campaigned in 1980 on a platform of deregulation, established a Presidential Task Force on Regulatory Relief chaired by Vice President George Bush. In early 1981 the task force selected the depressed automobile industry for priority consideration.

President Reagan appointed Drew Lewis to the Cabinet as secretary of transportation. Lewis was formerly lieutenant governor of Pennsylvania, where he was known as a conservative Republican with strong ties to the business community. Lewis is regarded as a smart politician who played a central organizational role in Reagan's 1980 election campaign. One of Lewis's first assignments in the new administration was to chair a Cabinet-level task force charged with the mission of helping the ailing auto industry. Lewis made several trips to Detroit in early 1981 to hear the needs of the automakers and explain the administration's commitment to regulatory relief.

At one meeting, Drew Lewis and White House counselor Ed Meese heard Roger Smith's case for help from Washington. At another meeting in Detroit, Lewis and Smith talked deregulation. Smith explained, "We showed Lewis a list of 1,500 regulations on the government schedules and especially those that aren't cost-effective."[7] One of the more prominent items on this list was the passive-restraint rule.

Meanwhile, David Stockman emerged as OMB director and as the administration's principal economic policy maker. In his famous "economic Dunkirk" memorandum for President Reagan, Stockman fashioned a series of tax cuts, spending reductions, and deregulatory moves designed to stimulate a supply-side expansion of the economy. The memorandum targeted the passive-restraint rule for repeal as part of a comprehensive regulatory relief package.[8]

On Capitol Hill the depression in the automobile industry had weakened political support for auto regulations of all kinds. In January 1981 Senator John Danforth (R–Mo.), an air bag advocate, held hearings on "Government Regulations Affecting the U.S. Automobile Industry" before his Subcommittee on Surface Transportation.[9] Representatives from GM, Ford, and Chrysler were asked to recommend what actions Washington could take to relieve the industry of burdensome regulations. Economist Larry White of the Graduate School of Business Administration at New York University, an expert on the auto industry, was asked to make an independent assessment of what regulatory relief measures would help the industry.

Testimony at the Danforth hearings revealed that the industry desired numerous actions, such as relaxation of NHTSA's bumper standards, weakening of EPA's emission rules for trucks, and relaxation of EPA's national ambient air quality standard for ozone. But the industry's top priority was elimination of NHTSA's passive-restraint requirements, which were due to take effect for large cars on September 1, 1981.[10] At a minimum, the domestic manufacturers requested delay of the passive-restraint requirements and a reversal of the compliance schedule so that small cars would be covered before large cars.

Professor White echoed much of the industry's testimony, calling for repeal of mandatory passive restraints. This is safety equipment, he argued, that consumers "would not voluntarily buy."[11] He also argued that the 1977 rule devised by former Secretary Adams was an indirect benefit to foreign producers. In particular, the compliance schedule was competitively favorable to importers of small cars whose products were not covered until model year 1984 (whereas large cars were covered beginning model year 1982).[12]

A Reprieve from Regulation

In early February 1981 Secretary of Transportation Drew Lewis proposed a one-year delay of the passive-restraint requirements. Citing estimates by GM and Ford, Lewis noted that the delay would allow the domestic industry to defer $40 million in capital investments and would save consumers $50 to $70 million in equipment costs.[13] When this proposal was adopted in final form on April 9, 1981, Lewis also announced that a comprehensive review of NHTSA regulations was underway. The explicit aim of the review was to reduce regulatory burdens on the troubled auto industry.[14] As part of this review Secretary Lewis reopened the passive-restraint rule making to consider repeal of the entire standard.[15]

The Lewis announcement led to delay or cancellation of the few air bag programs that were still in place. GM announced that "its inflatable restraint program has been terminated."[16] Mercedes, which was already selling cars with air bags throughout Europe, postponed plans to introduce air bags in the U.S. market.[17] BMW was planning to install air bags in three of its four 1983 models but announced delay until the 1984 model year due to regulatory uncertainty.[18] For the few surviving air bag suppliers, the Lewis announcement was a business nightmare.[19]

Recruiting an Expert Deregulator

Lewis elected not to resolve the passive-restraint issue personally. Instead, he delegated the controversial matter to the Reagan administration's choice to run NHTSA, Raymond Peck. In the 1960s Peck had been a Republican activist in New York State, but his legal career drew him to Washington, where in the 1970s he litigated on behalf of the coal industry against environmentalists. Peck was actually employed by the Department of the Interior when he was asked by colleagues to work for Martin Anderson and Ed Meese in the "policy arm" of President Reagan's transition team.[20]

In the process of doing "90-minute cram sessions" for the incoming Cabinet on emerging domestic issues, Peck caught the eye of Darrell Trent, who was Lewis's assistant secretary at the Department of Transportation. Lewis was persuaded by Trent that Peck was a good attorney with extensive experience in administrative law. Everyone agreed that the administration's regulatory relief program would require that NHTSA be run by a strong lawyer.[21]

The choice of Peck was also compatible with an early Reagan administration strategy: appoint leaders of regulatory agencies who are not imbued with a proregulation philosophy. In particular, Reagan avoided picking regulators with strong professional ties to the constituencies of federal health and safety agencies. Hence, Thorne Auchter of OSHA had little background in occupational safety and health, Ann Gorsuch of EPA had few ties to environmental groups, and Raymond Peck had no ties to the auto safety subculture.

Peck recalls only one meeting with Secretary Lewis where the passive-restraint issue was discussed in any detail. Peck explains: "Lewis told me that the President knows nothing about air bags. The Vice President knows nothing about air bags. You [Peck] are going to be the Administration's air bag expert. It is your issue and I can sell whatever decision you come to."[22] While this statement contradicts a widespread suspicion in Washington that Peck was ordered by Lewis from the start to rescind the passive-restraint standard, Peck knew which outcome his superiors in the administration would prefer—assuming it could be defended.

OMB's New Role

When the Reagan administration took office, the Council on Wage and Price Stability was abolished and its regulatory review functions transferred to the Office of Management and Budget. An executive order was issued that established a benefit-cost test for proposed regulations, and a strict clearance process was erected that required agencies to win OMB approval before a new regulation could be issued.[23]

OMB Director David Stockman delegated regulatory matters to the new Office of Information and Regulatory Affairs, which was run by economist James Miller. Since Miller also served as executive director of the Task Force, the efforts of OMB were easily coordinated with the actions of the Presidential Task Force on Regulatory Relief. Vice President George Bush was given broad authority to resolve disputes.

Miller pressured agencies to take a hard look at both regulatory proposals and rules in place to determine what was really needed. One OMB economist recalls that this pressure led to some "acrimonious discussions" with career officials in regulatory agencies.[24] The pressure for deregulation at NHTSA was enormous since all of the key White House players (Bush, Stockman, and Miller) were supportive of Lewis's campaign to provide regulatory relief to the auto industry. OMB analyst Diane Steed went to NHTSA—first as acting administrator and then as deputy administrator—and helped implement the regulatory relief program.

Within six months President Reagan nominated Miller to be a commissioner of the Federal Trade Commission. Miller was later succeeded at OMB by Christopher DeMuth, a scholar from Harvard's Kennedy School of Government who was known as an advocate of deregulation.[25] DeMuth did not arrive at OMB until September 1981, a few weeks prior to when Peck's final proposal was submitted for OMB approval.

The Insurance Industy Mobilizes

The insurance industry, led by Chairman Archibald Boe of Allstate Insurance Company, publicly supported mandatory passive restraints throughout the 1970s. Allstate was the first insurance carrier to offer a 30 percent discount on first-party medical coverage for air bag–equipped cars meeting NHTSA standards. The Insurance Institute for Highway Safety was a prolific source of pro–air bag comments to NHTSA. The President of IIHS was the former

NHSB Director, Dr. William Haddon, Jr. For years he had been the most articulate scientific spokesman for the injury-mitigation potential of passive restraints.

As the Reagan administration began to reconsider the passive-restraint standard, representatives of the insurance industry realized that they would have to play hardball if they were serious about stimulating the availability of automatic-protection technologies. The industry surprised many observers by taking two fairly aggressive steps.

First, a consortium of companies (Allstate, Kemper, Nationwide, State Farm, and Traveler's) commissioned Yale professor William Nordhaus to conduct a benefit-cost analysis of mandatory passive restraints. Nordhaus was an eminent economist with Washington experience as a member of the Council of Economic Advisors and chairman of the Carter administration's Regulatory Analysis Review Group.

Second, State Farm hired Arnold and Porter, a blue-chip Washington law firm, to monitor the legality of DOT's review of the passive-restraint standard. One of Arnold and Porter's first assignments was to advise State Farm on whether financial troubles within the domestic auto industry were a legally permissible basis for rescinding a safety standard under the terms of the 1966 Vehicle Safety Act.

The Nordhaus analysis, which was submitted to both NHTSA and Congress, concluded that the passive-restraint rule would generate favorable net benefits of $10 billion for model years 1982–1985. The major assumptions of Nordhaus's study follow:

- Almost all manufacturers would comply with the standard by installing automatic belt designs.
- Automatic belt designs would increase the rate of belt use from 12 to 57 percent.
- The new belt designs would reduce an occupant's risk of fatality and serious injury by 45 percent.
- The incremental cost of these new belt designs would be $60 per vehicle.
- The present value of insurance savings from reduced medical costs would be $30 per vehicle.

Nordhaus went further and predicted that delay of the passive-restraint standard would have a miniscule effect on the overall financial health of the troubled domestic auto industry.[26]

Meanwhile, State Farm's general counsel, Donald McHugh, released a written legal opinion by Arnold and Porter that concluded that financial troubles in the auto industry are not a legitimate reason to delay or rescind an otherwise cost-effective safety standard.[27] The opinion was presented to both NHTSA and to congressional committees interested in the passive-restraint issue. This move let Secretary Lewis and NHTSA officials know that the outcome of the upcoming rule making might be challenged in federal court if the letter and intent of the 1966 Vehicle Safety Act were not carried out.

The National Coalition to Reduce Car Crash Injuries

What became known as the "air bag lobby," the National Coalition to Reduce Car Crash Injuries, was a Washington-based group that served as a coordinating mechanism for the activities of consumer groups, insurers, and medical groups. The Coalition and its task groups met regularly to exchange intelligence and discuss strategy. The key body was an informal steering committee which included Joan Claybrook, James Fitzpatrick (Arnold and Porter), Charles Bruse (Allstate), Charles Taylor (NAII), Jeanne McGowan (AIA), Bill Haddon and Brian O'Neill (IIHS), Clarence Ditlow (Nader's Center for Auto Safety), Charles Hurley (National Safety Council), and William Steponkus (representing USAA).

The Coalition per se was not as pivotal as its component parts, but it did serve some significant roles such as:

- mailings to editors of the editorial pages of the several hundred largest newspapers in the country;
- preparation and dissemination of information kits with extensive information about air bags; and
- mobilization of the component groups for participation in executive and congressional branch hearings.

Given the dismal economic situation in the early 1980s, the Coalition had to work hard in Washington to keep the air bag on the nation's political agenda.

A Policy of Persuasion

One of Peck's first moves at NHTSA was to begin planning a large-scale safety belt use educational effort. Information about safety belts was to be distributed throughout the nation to sell the belt use message to opinion leaders and community activists. The kick-off of the campaign was delayed several months due to planning complications and Peck's inability to meet with Lewis, who was preoccupied with the air traffic controller strike. Peck ultimately requested $10 million in both fiscal years 1982 and 1983 for this effort. This initiative was announced prior to Peck's final decision on automatic restraints and was intended to go forward regardless of the outcome of the passive-restraint rule making.[28]

Peck's educational initiative drew support from Charles Pulley, President of the American Seat Belt Council. Pulley was convinced that a sustained educational effort could raise voluntary rates of belt use from 10 to 35 percent—at least based on the experience of England.[29] Education also tended to pave the way for mandatory belt use laws by reducing public opposition to belts and fostering public awareness of safety concerns.[30]

In late April 1981 Secretary Lewis referred to Peck's initiative in a discussion with a UPI reporter. Lewis stated that, "If we're able to encourage people to use those belts that are already in the car . . . we'd probably save more lives through that than any compulsory passive restraint requirement."[31] This statement was seized upon by Congressman Timothy Wirth (D–Colo.) and air bag advocates as evidence that Lewis had prejudged the passive-restraint issue prior to completion of rule making.[32] Yet Lewis made clear to Wirth in a letter that the belt use education program was to go forward regardless of how the pending passive-restraint issue was resolved.[33]

Peck's belt use campaign received significant support from the auto industry and the White House. President Reagan and top auto executives announced a new "Get It Together" campaign at a White House press conference in 1981. The slogan was inserted as a sticker on the windows of every new car produced by Ford.[34] The entire effort was executed in spite of complaints by Joan Claybrook and some insurance industry representatives that education was a waste of time and money. Charles Hurley of the National Safety Council expressed the prevailing sentiment among safety leaders: "Peck's campaign was all buttons, lollipops, and balloons. There was no science behind it."[35]

Apathy on the Hill

The chairman of the House Subcommittee on Consumer Protection was Timothy Wirth (D–Colo.), a young liberal Democrat with a strong record in favor of consumer safety regulation and a brash and ambitious personality. Wirth believed that the human carnage and wasted resources from motor vehicle crashes demanded aggressive regulatory responses, such as mandatory automatic crash protection. In response to DOT's delay of mandatory passive restraints, Wirth called for hearings before his subcommittee in late April 1981.[36]

The Wirth hearings were purportedly called to hear testimony on three legislative proposals. A bill authored by Wirth called for the five largest manufacturers to install passive restraints in small and mid-sized cars in model year 1984 and in large cars in 1985. Smaller manufacturers would remain covered by the 1977 Adams plan.[37] Representative Jim Collins (R–Tex.), a conservative, sponsored a bill to rescind the automatic restraint standards.[38] And a coalition of sixteen legislators (including Congressmen Dingell and Broyhill) cosponsored a bill to delay mandatory passive restraints until model year 1984 for cars of all sizes.[39]

Since the Congress had come within three votes in the House of passing Senator Warner's compromise rule in December 1980, Wirth was looking toward another effort to save the passive-restraint rule through legislation.[40] The politics of regulation proved, however, to be more difficult in the 97th Congress than they were in the 96th Congress.

For starters, there was still John Dingell (D–Mich.), who now became chairman of the powerful House Committee on Energy and Commerce to which Wirth's subcommittee must report. Dingell, whose district includes suburban Detroit, was an adamant opponent of air bags and a frequent ally of the domestic auto industry on regulatory matters. When John Moss (D–Calif.) retired from the House in 1978, the air bag lost its most influential supporter in the House, and Dingell, left relatively unopposed, assumed a dominant position.

More importantly, the conservative swing in the 1980 elections caused the Democratic party to suffer a net loss of 33 House seats to the Republicans. The freshman Republicans were far more antiregulation and free-market oriented than the Democrats they replaced. Since the passive-restraint issue had become a sort of symbol of paternalistic Washington bureaucracy, it was unlikely that these new House members would become supporters of a regulatory initiative by Wirth.

Opposition to mandatory passive restraints was also greater in the Senate, even though the Warner compromise had passed there easily in 1980. The November elections saw the Democrats lose control of the Senate to the Republicans, and many of the new Republican senators professed strong antiregulation ideologies. At the same time, auto safety advocates witnessed the demise of their heroes in the Senate: Abraham Ribicoff (D–Conn.) retired, Warren Magnuson (D–Wash.) was defeated by moderate Slade Gorton (R–Wash.), and Gaylord Nelson (D–Wis.) was defeated by conservative Robert Kasten (R–Wis.).

Although some Republican senators supported air bags (e.g., Warner, Packwood, and Danforth), they were in a minority. Since the White House—especially Stockman at OMB—was already pushing for repeal of the passive-restraint rule, the Republican senators who supported the new technologies found themselves in a difficult position. It is not easy to oppose an incoming President on an issue that the White House sets as a first-year priority.

As a result, few senators and House members were inclined to be vocal and energetic about supporting passive restraints. With the exception of Wirth in the House and Danforth in the Senate, legislative interest in passive restraints during 1981–1982 was slight. By the end of Wirth's hearings, it was apparent to everyone that Congress was unlikely to act on the passive-restraint issue. In effect, DOT had a free hand to implement Reagan's campaign promise of regulatory relief.

Getting Rid of a Regulation

From the very beginning of his tenure, it was clear that Ray Peck was looking for ways to remove regulations. During his confirmation hearings before the Senate, Peck expressed a philosophical preference that automatic restraints be made available to consumers as options rather than as mandated safety features. This position drew a skeptical response from Senator Danforth (R–Mo.), who said that "it is very doubtful that the notion of voluntary passive restraints . . . is going to get the job done."[41] Peck was confirmed easily despite the objections of some consumer safety advocates.

In August 1981 Peck held public hearings about whether the passive-restraint standard should be revised or rescinded. Air bag advocates emphasized that the market trend to small, less crashworthy cars highlighted the importance of retaining the automatic restraint standard. Insurance industry spokespersons testified that no evidence in the record supported rescission of the rule while much real-world experience with automatic belts and air bags buttressed the case for regulation.[42] The automobile manufacturers uniformly called for rescission of the standard, citing excessive costs and public disapproval of automatic belts. They encouraged NHTSA to pursue public education about belt use and mandatory seat belt use laws.[43]

Peck claims that he tried to give the sensitive issue back to Secretary Lewis, but "he wouldn't touch it." Apparently Lewis had become preoccupied with defending his path-breaking decision to fire the striking air traffic controllers. "He did not," Peck recalls, "want to fight another controversial issue."[44] Instead, Peck took a briefcase of materials on a trip to Denver, where he patiently sorted through the evidence and testimony.

Peck decided, to the surprise of no one in the Reagan administration, that the passive-restraint standard should be rescinded. He worked diligently with NHTSA officials to prepare the rationale for this action. Worried about the possibility of unfavorable leaks to the media, Peck submitted the proposal to OMB's DeMuth with a request for a quick turnaround.

Due to the political sensitivity of this matter, the communications between DeMuth and Peck were shielded from career officials at OMB and NHTSA.[45] While this arrangement allowed for more candid discussions, it created much suspicion and resentment among advocates of mandatory passive restraints.[46]

DeMuth shared David Stockman's view that a mandatory passive-restraint standard was bad public policy. While passive restraints are "a good buy for some consumers, they are a bad buy for others."[47] The government "shouldn't force the purchase on everyone."[48] Moreover, it is rare that a technology is so promising that it should increase from 0 to 100 percent of the market immediately. DeMuth felt that technological diffusion works best as "a gradual learning process" where experience is gained from limited use and the precise extent of consumer demand is clarified over time.[49]

Peck's rescission proposal was the first major regulatory issue to hit OMB during DeMuth's tenure. As DeMuth recalls, "I made a big mistake by not demanding that Peck give us adequate time to do a careful review of the rationale for rescission."[50] DeMuth and his staff performed instead a hurried late night review of the package, convinced that the policy call was correct even though the written rationale "seemed oddly reasoned yet tolerable."[51] DeMuth recommended that Stockman approve the proposal and Stockman accepted DeMuth's advice without careful consideration. Hence, Peck got the quick OMB approval that he had sought.

Peck's Reasoning

In November 1981 Peck announced his decision to rescind the passive-restraint standard.[52] At a press conference announcing the decision, Peck acknowledged that "there is great disappointment, great frustration, and great chagrin" among NHTSA scientists and engineers.[53] Indeed, Peck's decision was apparently made in contradiction to the unanimous view of senior NHTSA officials that the standard should be retained in some form.[54]

In support of his decision, Peck made three key findings. First, he noted that—contrary to DOT's projections in 1977—almost all cars subject to the standard would be equipped with passive belts that are easy to detach, not air bags. Second, he concluded that detachable automatic belts were not likely to result in substantial increases in belt-wearing rates. Finally, he predicted that the public was likely to protest an expensive regulation that offers few safety benefits.[55]

While Peck concluded that rescission of the rule was appropriate, he also noted that "the agency plans to undertake new steps to promote the continued development and production of air bags."[56] On the same day of this rescission order, Peck called Vice Presidents Herbert Misch of Ford and Betsy Ancker-Johnson of GM to request that their companies participate in a voluntary air bag demonstration. The phone calls were followed by written letters to the leadership of each major automobile manufacturer.[57]

Although career officials at NHTSA disagreed with Peck's decision and knew it would be controversial, they did not necessarily believe it would be overturned. NHTSA's Mike Finkelstein, for example, recalls that "I thought Peck was vulnerable on air bags but I thought he would win in the end."[58] NHTSA's regulatory analysis accompanying Peck's decision contained no serious analysis of air bags as a technological alternative. As Peck explained, "We were so focused on the rule as a performance standard that we never thought of an air bag–only rule."[59]

Peck's decision was applauded in Detroit, but it became an immediate target of legal challenges in the U.S. Court of Appeals for the District of Columbia by State Farm Insurance Company and the National Association of Independent Insurers. These petitioners were supported by the American Public Health Association, the Epilepsy Foundation, and the American College of Preventive Medicine. Legal briefs in support of Peck's decision were filed by the Motor Vehicle Manufacturers Association, Pacific Legal Foundation, Consumer Alert, and the Automobile Importers of America.

As soon as Peck rescinded the passive-restraint standard, he lost whatever credibility he had with most career NHTSA officials and air bag advocates.[60] His relations with Congress were also strained, though this was to some extent an interpersonal problem. According to one of his key aides at NHTSA, "Peck had a combative personality that got him into trouble." He "loved an argument" and was often "his own worst enemy."[61] Peck entered the litigation on the passive-restraint issue with supreme confidence that he would win. As one aide explained, "Peck considered himself an expert on the law of deregulation—indeed he came to NHTSA with a mission and that mission was to get rid of regulations."[62]

State Farm Takes the Lead

The decision of the insurance industry to sue the Reagan administration on passive restraints was not a foregone conclusion. Reagan was a popular President and had received substantial support in his campaign from people associated with the insurance industry. The leadership of the insurance industry was notoriously conservative and was expecting to need help from the Reagan administration on a variety of public policy issues.

The logical choice to lead in the litigation was thought to be Allstate. After all, Allstate had been the leading proponent of air bags in the industry throughout the 1970s. Yet when Peck rescinded the passive-restraint rule, Allstate's top management—which had begun to moderate its views—was willing to sue NHTSA only if some other company took the lead.[63]

The buck ultimately stopped at the desk of the president of State Farm Insurance Company, Ed Rust, Sr. Historically, Rust had been quite enthusiastic about the air bag, but he had rarely taken the lead on this issue within the industry. After hearing a briefing on Peck's decision, Rust—who is described as "a man of few words"—said "yes" to the lawsuit.[64] The result was that one of D.C.'s best law firms (Arnold and Porter) and one of D.C.'s best litigators (James Fitzpatrick) were commissioned to battle Peck's attorneys in federal court.

If the insurance industry had not sued NHTSA, Ralph Nader would surely have done so. And Nader has access to a fine public interest litigator named Alan Morrison. The difference was that Arnold and Porter had access to much greater resources than Nader, and more importantly, the conservative Burger Court might be inclined to take more seriously the arguments of Arnold and Porter than the arguments of a Nader group. As events transpired, Peck's attorneys were forced to fight both Fitzpatrick and Morrison before the liberal D.C. Circuit Court of Appeals—before the issue got to the Supreme Court.

Auto Safety Back in the Media

In the 1970s auto safety never achieved the degree of public attention it had acquired in the 1960s through the efforts of Nader and congressional operators. In January 1982 a new dynamic caught the media's interest and again placed the auto safety issue in the nation's consciousness. The Insurance Institute for Highway Safety published data suggesting that Japanese cars were the least safe of the small cars available to buyers. This study may have tapped a latent nationalistic sentiment at a time when Detroit was in deep economic trouble due to imports and the deep recession.

Insurance executive General Robert McDermott held a press conference in Washington to publicize the IIHS results, which attracted 110 reporters. That night McDermott went on the "MacNeil/ Lehrer NewsHour," and the next day he met with the trade press in New York. The message about unsafe Japanese cars was heard by millions of Americans in a direct and coherent fashion.

A startling result of this media splash was a decision by General Motors to use the IIHS data in its marketing campaigns and annual reports. In the ensuing years Americans were exposed to a sustained communications initiative that sought to put safety in the public's consciousness. As we shall see, the increased public awareness of auto safety would prove to be a key factor in the ultimate resolution of this controversy.

Taxes Instead of Regulation?

Senator John Danforth (R–Mo.) is an Episcopalian minister, a Yale-trained lawyer, and a son of aristocracy. Claybrook describes him as "a bold politician with a quiet exterior and measured personality."[65] Danforth was by far the air bag's most persistent supporter on the Hill.

When Peck rescinded the passive-restraint requirements, Danforth was dismayed. He immediately introduced legislation to restore the requirements, but this bill generated little interest in the Congress.[66] Rather than abandon the passive-restraint issue, Danforth searched for an alternative legislative approach that might be more acceptable in an antiregulation political environment. Although litigation against Peck was underway, this tactic could take years to reach any resolution and Peck's decision might ultimately be upheld.

In late November 1981, Danforth introduced a new bill that called for tax incentives to encourage air bag installation. Carmakers would be permitted to claim a $300 tax credit for each 1984 car equipped with air bags, and a $300 excise tax would be levied on each car sold without air bags. No tax incentives would be offered for automatic belts. This proposal capitalized on Danforth's membership on the Senate Finance Committee, which had jurisdiction over the bill. Danforth persuaded Senator Robert Packwood (R–Oreg.), chairman of the subcommittee on Taxation and Debt Management, to schedule hearings on the bill for late January 1982. The tax incentive plan was particularly attractive because it could be passed as a rider to any tax or debt management bill.[67]

While Danforth's bill drew support from insurers and consumer advocates, it was opposed by the Treasury Department.[68] J. Gregory Ballantine, deputy assistant secretary of the Treasury Department, argued that it was inappropriate to shift the costs of air bags from the consumers of cars to all taxpayers. Moreover, Ballantine said, there was a risk—especially in the short run—that the bill would fail to offer sufficient incentives to bring air bags to the marketplace. The result would be the equivalent of a $300 excise tax on new cars that would further reduce sales and aggravate the financial troubles of the domestic industry.[69]

Administrator Peck also testified against the tax-incentive plan. He preferred to pursue an air bag demonstration program, perhaps one similar to the package negotiated by Secretary Coleman in December 1976.[70] Although Ford Motor Company expressed a willingness to participate in a cooperative demonstration program, the other major car manufacturers remained cool to the idea. GM Chairman Roger Smith made it clear to Peck that GM would make no more investments in air bags.[71] Since Peck had already rescinded the passive-restraint rule, he had no direct leverage—as former Secretary Coleman did—to induce car companies to come to the bargaining table. If Peck had tried prior to rescission to negotiate a demonstration program, he probably would have faced difficult lawsuits.[72] In the final analysis, Danforth's plan went nowhere and Peck struggled to establish an air bag demonstration.

Suing Detroit

As the Reagan administration's regulatory relief program dimmed the prospects for air bags, several creative lawyers began to look into the potential of tort suits as a technique to compensate crash victims and keep pressure for air bags on the auto industry. Aside from several articles in academic journals, the notion of crash victims suing car manufacturers for failure to install air bags was not taken seriously in the 1970s. The leaders of this effort were Joan Claybrook, Stephen Teret (an academic at Johns Hopkins University's School of Public Health), and Andrew Hricko (an attorney at the Insurance Institute for Highway Safety).

As early as 1977 Hricko suggested in a published article that the entire auto industry could be held liable for failure to install air bags in cars.[73] Hricko pointed to the holding in the T.J. Hooper case. This case arose from the failure of the tugboat industry to equip its vessels with radios capable of receiving weather reports. Judge Learned Hand ruled on behalf of the victims of a sunken barge whose damages may have been prevented if vessels had been equipped with radios.[74] Hricko's analogy of this case to the air bag went unnoticed by the plaintiff's bar until the early 1980s.

Stephen Teret publicized Hricko's thesis in a July 1982 article in *Trial*, the magazine of the Association of Trial Lawyers of America. Teret and his co-author concluded, "The message of large verdicts for the failure to make air bags can be loudly heard by automobile manufacturers, and has the potential for being more effective than the attempts to regulate over the past dozen years."[75] Following publication of this article, Teret reported that cases against the automakers began being filed across the country.

In May 1982 Claybrook helped organize a meeting of the Trial Lawyers for Public Justice to train activist attorneys on how to initiate air bag lawsuits.[76] Claybrook also signed up as a rebuttal witness for a woman in Alabama who was severely injured in a Ford Pinto. The woman was prepared to argue that her injuries would have been much less serious if Ford had installed air bags in her car. Ford settled the case before it went to trial for $1.8 million.[77]

Legal experts at Ford characterize this description of the Alabama case as a "canard." The crash victim was an attractive young college student (a former cheerleader) who occupied the passenger side when a Camaro struck the Pinto at about a 45 degree angle. The reasons for the settlement are complex but, insist Ford officials, have nothing to do with the lack of an air bag.[78] Indeed, Ford and other manufacturers have won dismissals of more than thirty "air bag suits" on the grounds that their compliance with the occupant-protection standards of the 1966 Safety Act preempts such tort suits. Plaintiff attorneys, however, have won the preemption issue in a recent Massachusetts case that is now being appealed. In no case so far have automakers lost a tort suit on the grounds that vehicles were not equipped with air bags, although no such case has yet gone to a verdict.[79]

It is difficult to assess what effect, if any, these tort suits have had on the thinking of managers in the auto industry. Inasmuch as the introduction of air bags itself was perceived by industry to pose additional liability exposure, it is not clear what corporate policy would minimize a firm's overall liability exposure. According to Roger Maugh of Ford Motor Company, "These lawsuits were not a critical factor in our thinking."[80] David Martin of GM believes that liability risk was never a "critical factor" in his company's policy, although the litigious environment did enhance GM's incentive to "refine" the air bag system.[81] The more serious litigation strategy was directed not at Detroit but at the Reagan administration.

"Not One Iota of Evidence"

The lawsuits against Peck's rescission order were filed by insurers in the U.S. Circuit Court of Appeals for the District of Columbia—the "D.C. Circuit." Of the twelve federal appeals courts located throughout the nation, the D.C. Circuit is responsible for the majority of administrative law cases. In the 1970s this circuit developed a reputation for activism in review of agency decisions—ordering more public hearings, finding "arbitrary and capricious" agency behavior, and second-guessing administrative interpretations of legislative mandates. The more conservative Supreme Court, under the leadership of Warren Burger, was frequently reversing D.C. Circuit decisions and issuing opinions that reprimanded D.C. Circuit judges for excessive intrusion into agency discretion.

When the passive-restraint case came up for oral argument in March 1982, the composition of the three-judge panel seemed favorable to air bag advocates. The senior judge was David L. Bazelon, a veteran liberal of the D.C. Circuit who was not reluctant to scrutinize agency procedures and decisions. The junior judges were two liberals who had been appointed by President Carter: Abner J. Mikva and Harry T. Edwards. Mikva was a former Democratic congressman from Illinois and Edwards a former law professor and relatively recent appointment to the court.

The legal case against Peck was not easy to make in that Congress had never passed a law requiring NHTSA to enact a passive-restraint rule. The 1966 Safety Act did say generally that both adoption and revocation of standards were subject to court review. Lawyers for the insurance industry urged the court to hold that Peck's decision was unlawful on the grounds that it was an "arbitrary and capricious" use of administrative discretion.

The Administrative Procedure Act authorized federal courts to review agency decisions to determine whether they are based on "substantial evidence" in the agency's rule-making record. Insurers stressed that the lifesaving benefits of air bags and automatic belts were not in doubt. If the panel found Peck's decision to be unlawful, the judges could order reinstatement of the standard, require further analysis of the issue, and/or compel NHTSA to disallow detachable passive belts.

Lawyers for NHTSA and the auto industry urged the court to defer to the expert judgment of NHTSA on the appropriateness of a mandatory passive-restraint standard. They argued furthermore that the courts should engage in no more intensive scrutiny of a decision to deregulate than a decision to regulate. If the agency's decision should be ruled unlawful, they urged the court to let the agency take corrective measures without judicial interference.

In June 1982, nine months after Peck's decision, the three-judge panel held that the rescission order was unlawful. The court ordered the agency to prepare and submit to the court within thirty days a schedule for completing analysis of questions raised in the court's opinion.[82] The court later ordered reinstatement of the standard and ordered it to go into effect for all new cars beginning with model year 1984.[83]

Writing on behalf of a unanimous panel, Judge Abner Mikva emphasized that there was "not one iota of evidence" to support Administrator Peck's determination that detachable automatic belts would not increase belt use.[84] Judges Mikva and Bazelon (with Judge Edwards dissenting only in part) also faulted Peck for not exploring certain alternatives to deregulation, such as disallowance of detachable passive belts and mandatory air bags. According to Judge Mikva's opinion, NHTSA had "wasted administrative and judicial resources" with "an expensive example of ineffective regulation" of "the worst kind."[85] In his concurring opinion, Judge Edwards wrote that Peck's decision "appears to be nothing more than a determined effort to achieve a particular result without regard to the facts at hand."[86]

Judge Mikva's opinion was perceived by Reagan administration officials as more than just a setback on passive restraints. Some of the language in the opinion appeared to create new legal barriers to implementation of the administration's regulatory relief program. For example, Mikva inferred congressional support of passive restraints from "legislative silence" (i.e., congressional inaction). Moreover, he seemed to create a greater evidential burden for acts of deregulation than for acts of regulation.[87] As a result, the U.S. Department of Justice—led by Solicitor General Rex Lee—decided to represent NHTSA in an appeal of the decision to the nation's highest Court.

The appeal to the Supreme Court was initiated in spite of efforts by Republican Senators John Danforth and Orrin Hatch to persuade the administration to accept the D.C. Circuit's order. The Republican Senate went so far as to pass an amendment to a supplemental appropriations bill that would have prohibited NHTSA from expending taxpayer funds on further legal appeals of the D.C. Circuit decision.[88] But the amendment was dropped by House-Senate conferees. In any case the amendment was only symbolic because the Motor Vehicle Manufacturers Association, the Automobile Importers Association, Consumer Alert, and the Pacific Legal Foundation had already petitioned the Supreme Court for a writ of certiorari.[89] While observers waited for word from the Burger Court, some efforts were being made by Peck to make air bags available.

Reviving an Old Idea

After rescinding the passive-restraint rule, Administrator Peck wrote the executives of the major car manufacturers seeking voluntary participation in an air bag demonstration program. Peck liked the idea that former Secretary Coleman had tried to implement in 1976. The responses were not enthusiastic, especially from GM Chairman Roger Smith and Chrysler Chairman Lee Iacocca. Instead of giving up, Peck followed up on an earlier idea of Ralph Nader's to have the federal government purchase a fleet of air bag–equipped cars for government use. This program would guarantee a minimum market for cars with air bags, and if successful, state and local governments and large commercial buyers might be persuaded to take similar steps.

Since government cars were purchased by the General Services Administration, Peck entered into negotiations with GSA Administrator Gerald P. Carmen. Ralph Nader helped Peck persuade Carmen to support the effort.[90] In late December 1982 NHTSA and GSA announced a plan to equip up to 5,000 cars with air bags as early as the 1985 model year.[91] The plan called for the devices to be installed as original equipment on the driver side. When the plan was announced, no car manufacturer had yet committed to being a supplier for the program. Barring a willing original equipment supplier, the plan called for cars to be retrofitted with the devices at NHTSA's expense.

Most auto manufacturers decided not to participate in the demonstration program. Christopher Kennedy of Chrysler Corporation, a major supplier of government cars, stated that "we don't think 5,000 cars would be enough to justify the research and development costs and the tooling costs."[92] The only positive reactions came from Roger Maugh, director of automotive safety at Ford Motor Company, who said that installation of air bags in a fleet of 1985 models might be attainable. Actually, Ford had revived its air bag program in 1981. According to Maugh, "We didn't want both marketing and technical problems at the same time; the GSA plan helped solve the initial marketing problem."[93]

Mercedes-Benz Takes the First Step

In January 1983 Peck's desire to make air bags available to the public was boosted by the decision of Mercedes-Benz of North America to offer driver-side air bags as an option on certain 1984 models.[94] The so-called "Supplemental Restraint System" (SRS) consisted of a standard lap/shoulder belt system supplemented by an air bag and knee bolster on the driver's side and an Emergency Tensioning Retractor attached to the right-front passenger's belts. The word "supplemental" was a key breakthrough in thinking because it meant that belts were considered complementary rather than competitive with air bags.

The company's marketing plan called for the new system to be installed in 5,000 1984 vehicles. If customer reactions proved favorable, Mercedes was prepared to make the option available on all its 1985 and later models. The projected price for the option was $800 to $900 per car. The SRS had already become a very popular safety feature among European customers.[95]

While Ford and Mercedes-Benz were pushing forward with air bag development, some automakers appeared to be confident of a victory in the Supreme Court. The *Wall Street Journal* reported in February 1983 that development work on automatic belts had been suspended at Ford, GM, and Chrysler.[96] One engineer at Chrysler was told by upper-level management that "they were confident the Supreme Court would rule that [automatic] belts weren't needed."[97]

Departures from the Reagan Administration

In early 1983 Secretary Lewis resigned his Cabinet position to assume the chairmanship of Union Pacific Corporation. To replace him President Reagan nominated Elizabeth "Liddie" Dole, a former commissioner of the Federal Trade Commission and the wife of Senator and Majority Leader Robert Dole (R–Kans.). In contrast to the conservative Lewis, Dole was considered a moderate-to-liberal Republican with a strong commitment to consumer protection.

Dole's appointment was followed by the resignation of Peck from the leadership of NHTSA. This resignation was not a complete surprise, considering that Peck's relations with several key legislators were terrible. In early 1983, in fact, Peck was thrown out of a congressional appropriations hearing by Congressman Adam Benjamin (D–Ind.).[98] Speculation in Detroit was that Dole pushed Peck out in order to chart a new course in auto safety regulation.[99] Career officials at NHTSA reached the same conclusion.[100]

Peck's resignation was greeted with some disappointment in Detroit. "He's the only friend we've had there in a long time," one auto industry official said.[101] At consumer safety organizations such as Ralph Nader's Center for Auto Safety, the resignation of Peck was not a cause for sorrow. In addition, the *Automotive News* quoted several NHTSA staffers as saying that "there was jubilation in the NHTSA hallways."[102]

In a *Washington Post* interview soon after her Senate confirmation, Secretary Dole described the air bag as "a good safety device."[103] She also discussed several approaches she intended to explore to promote the lifesaving technology. When asked to comment on the upcoming Supreme Court decision, she declined.

The Burger Court Rules

When the Supreme Court decided in late 1982 to hear the solicitor general's appeal of NHTSA's case, the D.C. Circuit responded by withdrawing the court-imposed requirement that passive restraints be installed in all 1984 cars. Despite Judge Mikva's decision, the supporters of deregulation had good reason to be hopeful.

The fact that the Court agreed to hear the solicitor general's appeal was a hint that at least four justices—the minimum required to accept an appeal—felt that it was inadvisable to let the Mikva opinion stand without careful review. Moreover, NHTSA was to be represented in oral argument before the Court by Solicitor General Rex Lee, an eminent attorney with an impressive track record in appeals to the Burger Court. Lee's position was also buttressed at oral argument by veteran attorney Lloyd Cutler, who represented the Motor Vehicle Manufacturers Association. The composition of the highest court was also more conservative than the D.C. Circuit, especially after President Reagan's appointment of Justice

Sandra Day O'Connor. At oral argument, Lee and Cutler pointed to several legal interpretations in Judge Mikva's opinion that were inconsistent with prior decisions of the Burger Court.

In June 1983 the Burger Court surprised virtually everyone by holding that Administrator Peck's decision was indeed unlawful.[104] Writing for a unanimous court, Justice Byron White stated that it was "arbitrary and capricious" of Administrator Peck to rescind the entire standard without first examining the injury-mitigation potential of air bags and nondetachable automatic belts. Justice White added that auto manufacturers had waged "the regulatory equivalent of war against air bags and lost."[105]

A majority of the Court (Justices White, Harry Blackmun, John Paul Stevens, Thurgood Marshall, and William Brennan) also found that Peck did not undertake a balanced analysis of the lifesaving potential of detachable automatic belts. Justice Stevens had been particularly vocal at oral argument about Peck's illogical analysis of detachable automatic belts. Justice Stevens told Solicitor General Lee: "This is a matter of common sense. If it's automatically attached, then you have to do something affirmative to detach it. Isn't it perfectly obvious that more people would use it? Even a one percent increase in belt use would save 100 lives."[106] The Court also ordered NHTSA to reconsider the benefits of detachable automatic belts in light of the Court's opinion.

The case was actually much closer than it may appear at first glance. Several clerks at the Court have said privately that the case was 4–4 on detachable automatic belts before Justice Blackmun made up his mind. If Justice Blackmun had not joined Justice White's opinion, the unanimous opinion on air bags might have been written by Justice Rehnquist instead of Justice White. The holding would have been the same, but the opinion might have read that the Reagan administration must simply come up with some decent arguments against air bags. When Justice Blackmun finally joined Justice White's opinion, the case turned into a big victory for the insurance industry.

In the final analysis, the Supreme Court's decision differed from the D.C. Circuit's decision not in the outcome but in some of the legal reasoning provided to support the reversal. Justice White stated that Judge Mikva erroneously relied upon unenacted legislative proposals as though they were legislative history and imposed an excessively stringent standard of review to acts of deregulation. Justice White's opinion did agree with Judge Mikva's observation that "NHTSA's analysis . . . of air bags was nonexistent."[107] With this statement, the Supreme Court returned the passive-restraint issue to America's national political agenda.

Conclusion

From the perspective of administrative management, it is difficult to defend Raymond Peck's handling of the passive-restraint issue. His performance was poor not necessarily because he sought to rescind the passive-restraint rule but because the evidence and reasoning he used as his rationale were woefully inadequate. By advancing a poorly reasoned decision, he left the Reagan administration vulnerable to charges that deregulation was a mere fulfillment of a campaign promise rather than a reasoned policy decision. If Peck had challenged both automatic belts and air bags as being economically unreasonable, he might have fared better in the Burger Court. In light of the complexity and sensitivity of this issue, Drew Lewis and David Stockman did not serve the Reagan administration well by refusing a personal role in the decision.

From the perspective of corporate strategy, the leadership of State Farm and other insurers was decisive and impressive. They saw that their commercial interests were not compatible with the Reagan administration's program of regulatory relief, even though they may have had some ideological sympathies with the Reagan administration. Their tactics followed a textbook Washington pattern: Solicit unimpeachable expert testimony and hire a blue-chip law firm to build the case for a powerful lawsuit. Though the car manufacturers and NHTSA may have been prepared to withstand a vigorous challenge from the likes of Ralph Nader and Joan Claybrook, they were not capable of defending deregulation when the nation's insurance industry was determined to play hardball. The insurers won the battle of the titans.

NHTSA's vulnerabilities in court were exacerbated by some ill-considered strategy by Roger Smith and Lee Iacocca. Given Peck's refusal to criticize air bag technology, NHTSA needed some concrete plan to make the device available to the public *without regulation*. By giving Peck the cold shoulder on his invitation to participate in a voluntary demonstration program, Smith and Iacocca tightened (perhaps inadvertently) the noose around Peck's neck. In any event, it is clear that Smith and Iacocca seriously underestimated the willingness and power of the insurance industry to block their ambitions for deregulation.

Endnotes

1. *Motor Vehicles Facts and Figures 1984*, U.S. MVMA, Detroit, Michigan (1985), p. 16.

2. Comments of William Nordhaus on Notice of Proposed Rulemaking on Federal Motor Vehicle Safety Standard: Occupant Crash Protection, U.S. Dept. of Transportation, NHTSA, 1981, p. 3.

3. *Statistical Abstract of the United States*, U.S. Dept. of Commerce, Washington, D.C., various years.

4. Comments of William Nordhaus, p. 4.

5. *Motor Vehicles Facts and Figures 1984*, p. 16.

6. Ibid.

7. Ralph Nader and William Taylor, *The Big Boys: Power and Position in American Business* (New York: Pantheon Books, 1986), p. 101.

8. Michael Wines, "Trying to Rescind the Automobile Air Bag Rule: A Case Study of the Headaches of Deregulation," *National Journal* (January 1, 1983), pp. 8–9.

9. *Government Regulations Affecting the U.S. Automobile Industry*, Hearings before Subcommittee on Surface Transportation, Committee on Commerce, U.S. Senate, 97th Congress, First Session, January 28, 1981, pp. 2–3.

10. Testimony of David S. Potter, vice president, Public Affairs Group, General Motors Corporation, ibid., p. 27.

11. Testimony of Lawrence J. White, Professor of Economics, New York University Graduate School of Business Administration, ibid., p. 18.

12. Ibid., p. 21.

13. *Federal Register*, 46:12033–12034 (1981).

14. "Actions to Help the U.S. Auto Industry," Office of the Press Secretary, The White House, April 6, 1981.

15. *Federal Register*, 46:21205–21208 (1981).

16. "General Motors Drops Efforts to Improve Air Bags," *New York Times* (April 30, 1981), p. A28.

17. "Mercedes Was Ready to Use Air Bags in U.S. Cars," *Status Report: Highway Loss Reduction* (November 5, 1981), p. 3.

18. Letter of Hanns Weisbarth, vice president, Engineering, BMW of North America, Inc. to Congressman Timothy Wirth, May 18, 1981.

19. Statement of Ralph Rockow, vice president, Talley Industries, Inc., and chairman, Automotive Occupant Protection Association, in *Automatic Crash Protection Standards*, Hearings before Subcommittee on Telecommunications, Consumer Protection and Finance, House Committee on Energy and Commerce, 97th Congress, First Session, April 1981, pp. 207–209.

20. Interview of Raymond Peck, July 30, 1987.

21. Ibid.

22. Ibid.

23. W. Kip Viscusi, "Presidential Oversight: Controlling the Regulators," *Journal of Policy Analysis and Management* 2:160–163; 1983.

24. Interview of Thomas Hopkins, September 18, 1987.

25. Christopher DeMuth, "Constraining Regulatory Costs—Part I: The White House Review Programs," *Regulation* 4:13–26 (1980).

26. Comments of William Nordhaus on Notice of Proposed Rulemaking on FMVSS 208: Occupant Crash Protection, U.S. Dept. of Transportation, NHTSA, 1981.

27. Memorandum from Arnold and Porter to State Farm Mutual Automobile

Insurance Company, "Legal Issues Raised by NHTSA's Consideration of the General Economic Condition of the Automobile Industry in Amending the Passive Restraint Standard," April 27, 1981.

28. "Seat Belt Promotion Delayed," *Status Report: Highway Loss Reduction* (September 4, 1981), pp. 1–2.

29. Testimony of Charles Pulley, in *Automatic Crash Protection Standards,* Hearings before the Subcommittee on Telecommunications, Consumer Protection, and Finance, House Committee on Energy and Commerce, 97th Congress, First Session, April 1981, p. 211.

30. Motorists Information Inc., *Seat Belt Education Program: Post Advertising Test,* Summary Report, Lincorp Research, Inc., Southfield, MI, June 1977.

31. "Lewis Favors Seat Belts Over Compulsory Restraints," *Pittsburgh Press* (April 26, 1981), p. A-23.

32. Letter from Congressman Timothy Wirth to Secretary Drew Lewis, April 28, 1981.

33. Letter from Secretary Drew Lewis to Congressman Timothy Wirth, April 29, 1981.

34. Interview of Roger Maugh, July 8, 1987.

35. "Promoting Belt Use: Lessons from the Past," *Status Report: Highway Loss Reduction* (June 24, 1981), pp. 1–3; "Buckling Up: Regulators Stick to the Soft Sell," *Business Week* (October 25, 1982), p. 110c; interview with Charles Hurley, June 22, 1988.

36. *Automatic Crash Protection Standards,* 1981.

37. H.R. 3237, 97th Congress, First Session (April 10, 1981).

38. H.R. 3184, 97th Congress, First Session (April 9, 1981).

39. H.R. 3151, 97th Congress, First Session (April 8, 1981); "Dingell Studies Two-Year Delay," *Status Report: Highway Loss Reduction* (February 25, 1981), p. 2.

40. "Committee Leader Backs Automatic Restraints," *Status Report: Highway Loss Reduction* (May 31, 1981), p. 1.

41. "Danforth Unwilling to Scuttle Automatic Safety Rule," *Status Report: Highway Loss Reduction* (April 27, 1981), p. 4.

42. See, for example, Letter of William Haddon Jr., President, Insurance Institute for Highway Safety, to NHTSA Docket 74-14, Notice 22, FMVSS 208: Occupant Crash Protection, May 26, 1981.

43. See, for example, Comments of General Motors Corporation Regarding Passive Restraints (FMVSS 208), NHTSA Docket 74-14, Notice 22, May 26, 1981.

44. Interview of Raymond Peck, July 30, 1987.

45. Interview of Thomas Hopkins, September 18, 1987; interview of Michael Finkelstein, July 23, 1987.

46. Joan Claybrook and the Staff of Public Citizen, *Retreat from Safety: Reagan's Attack on America's Health* (New York: Pantheon Books, 1984), pp. xvii, xix–xxii.

47. Interview with Christopher DeMuth, July 28, 1988.

48. Ibid.

49. Ibid.

50. Ibid.

51. Ibid.
52. *Federal Register*, 46:53419–53429 (1981).
53. "NHTSA Abandons Automatic Restraint Standard," *Status Report: Highway Loss Reduction* (November 5, 1981), pp. 1–2.
54. Interview with Barry Felrice, July 23, 1987.
55. *Federal Register*, 46:53420–53424 (1981).
56. "NHTSA Abandons Automatic Restraint Standard," p. 3.
57. Interview with Raymond Peck, July 30, 1987.
58. Interview of Michael Finkelstein, July 23, 1987.
59. Interview of Raymond Peck, July 30, 1987.
60. Interview of Michael Finkelstein, July 23, 1987.
61. Interview of Barry Felrice, July 23, 1987.
62. Ibid.
63. Interview of Brian O'Neill, July 24, 1987.
64. Ibid.
65. Interview of Joan Claybrook, July 24, 1987.
66. Senator John Danforth, *Congressional Record* (October 26, 1981), pp. S12150–S12151
67. "Tax Incentives Proposed for Air Bag Installation," *Status Report: Highway Loss Reduction* (December 9, 1981), pp. 3–4.
68. "Tax Incentives for Air Bag Draw Support," *Status Report: Highway Loss Reduction* (February 17, 1982), pp. 4–5.
69. Statement of J. Gregory Ballentine, Deputy Assistant Secretary (Tax Analysis), Dept. of the Treasury, Hearings before Subcommittee on Taxation and Debt Management, Senate Finance Committee, March 2, 1982, mimeo.
70. Statement of Raymond A. Peck, Administrator, NHTSA, Hearings before Subcommittee on Taxation and Debt Management, Senate Finance Committee, March 2, 1982, mimeo.
71. Interview of Raymond Peck, July 30, 1987.
72. Ibid.
73. Andrew R. Hricko, "T.J. Hooper and the Air Bag," *Federation of Insurance Counsel Quarterly* 27 (1977).
74. *The T.J. Hooper*, 60 F.2d. (2nd Circuit 1932), p. 737.
75. Stephen Teret and Edward Downey, "Air Bag Litigation: Promoting Passenger Safety," *Trial* 18:93–99 (1982).
76. Interview of Joan Claybrook, July 24, 1987.
77. Ibid.
78. Interview of Don McGuigan, September 3, 1987.
79. William F. Doherty, "U.S. Court to Rule on Car Air Bag Suit," *Boston Globe* (December 6, 1987), p. 41.
80. Interview of Roger Maugh, July 8, 1987.
81. Interview of David Martin, September 3, 1987.
82. *State Farm* vs. *DOT*, 680 F.2d 206 (D.C. Circuit 1982).
83. " '84 Cars Must Have Automatic Restraints," *Status Report: Highway Loss Reduction* (August 12, 1982), pp. 1, 7.
84. "Court Rejects NHTSA Action in Restraints," *Status Report: Highway Loss Reduction* (June 9, 1982), p. 1.
85. Ibid.

86. Ibid.

87. "Active Judges and Passive Restraints," *Regulation* (July/August 1982), pp. 10–15; "Hot Air Bags," *Harper's* (December 1982), pp. 19–20.

88. Statement of Senator John Danforth, *Congressional Record* (August 10, 1982), p. S10058; statement of Senator Orrin Hatch, *Congressional Record* (July 12, 1982), p. S8062.

89. "Auto Makers Appeal Restraints Decision," *Status Report: Highway Loss Reduction* (September 2, 1982), p. 1.

90. Interview with Michael Finkelstein, July 23, 1987.

91. "Air Bag Demo Test Moves a Step Closer," *Automotive News* (December 20, 1982), p. 3.

92. "NHTSA Floats Trial Air Bags," *Washington Post* (January 8, 1983), p. D9.

93. Interview of Roger Maugh, July 8, 1987.

94. Letter from W.R.F. Bodack, President of Mercedes-Benz of North America, to NHTSA Administrator Raymond Peck, January 28, 1983.

95. "Mercedes Plans Air Bags in '84 Models," *Status Report: Highway Loss Reduction* (February 1, 1983), pp. 1, 7; "Mercedes to Introduce Air Bag Option in '84," *Automotive News* (February 7, 1983), p. 8.

96. "Auto Makers Drop Automatic Seat Belts Although High Court Ruling Still Awaited," *Wall Street Journal* (February 4, 1983), p. 10.

97. Ibid.

98. Interview of Barry Felrice, July 23, 1987.

99. "Peck Quits as NHTSA Chief," *Automotive News* (April 25, 1983), p. 1.

100. Interview of Barry Felrice, July 23, 1987; interview of Charles Livingston, July 24, 1987; interview of Michael Finkelstein, July 23, 1987.

101. "Peck Quits as NHTSA Chief," p. 1.

102. Ibid.

103. "Dole Touts Air Bags as U.S. Fights Them," *Washington Post* (April 10, 1983), p. A6.

104. *Motor Vehicle Manufacturers Association* vs. *State Farm Mutual Automobile Insurance Company*, 103 S.Ct. (1983), p. 2856.

105. Ibid., p. 2870.

106. "Supreme Court Is Told Restraint Rescission Violates DOT Mandate," *Status Report: Highway Loss Reduction* (May 12, 1983), p. 4.

107. Ibid., p. 2869.

Chapter 8

THE MASTER PLURALIST

Justice Byron White's opinion in the *State Farm* case left the Reagan administration with the proverbial hot potato. The insurance industry and consumer activists were determined to scrutinize each step the administration made on the passive-restraint issue. In addition, the leadership of the auto industry had good reason to be annoyed at how Ray Peck had bungled the initial attempt at deregulation. Yet, the Reagan White House and Stockman's OMB had absolutely no interest in reinstating a multibillion-dollar regulation fashioned by Joan Claybrook during the Carter administration.

Elizabeth Hanford Dole, President Reagan's second secretary of Transportation, fashioned the administration's response to the Supreme Court's 1983 decision. Her handling of the issue is likely to go down in American history as one of the more masterful resolutions of a seemingly unresolvable domestic political controversy. She harnessed the energies of political conflict that permeated the 1970s and rechanneled them toward a constructive safety program that has ultimately won praise from virtually everyone.

Although Elizabeth Dole emerges as the "master pluralist" in this saga, the point is not that her unique political talents were responsible for the breakthrough. Her handling of other issues at DOT was much less proficient (e.g., the process of selling Conrail), and she made her share of mistakes on occupant restraints. The key point is that the success of her plan arose from a shrewd recognition of how pluralism in America works. The Dole plan offered something for everyone at the same time as it assured progress toward the safety goals set by Congress almost twenty years earlier.

Some skeptics insist that some or all of the favorable outcomes of Dole's plan were fortuitous. The argue that Dole did not intend or foresee the dramatic safety advances that, as we shall see, her

169

decisions generated. One need not dispute these skeptics to praise Dole's plan as a model of democratic pluralism. After all, one of the essential propositions of pluralism is that the public interest arises unpredictably out of the competition of interest groups. Elizabeth Dole set in motion a new type of political competition that was directed at enhanced safety. The beauty of her decision is that she did not try to dictate which road to safety America would choose.

Power Struggle

The Supreme Court's decision on passive restraints did not compel the Reagan administration to take any particular rule-making action. Under the terms of Justice White's opinion for the Court, the passive-restraint rule could be reissued, amended, or recinded again with better justification. No deadlines for resolution of the matter were prescribed in White's opinion. The result of the Court's open-ended remand was a power struggle within the Reagan administration over what course of action would be taken and who would decide.

Because the Reagan administration had not yet nominated a person to succeed Raymond Peck, how the issue could be resolved at NHTSA was very unclear. Secretary of Transportation Elizabeth Dole said that "she welcomed the opportunity to review this matter under guidance provided by the Supreme Court."[1] She emphasized that "I have no higher priority than to insure safety."[2] Indeed, some auto industry officials feared Dole because they regarded her as a "closet Naderite."[3]

Some doubted whether Dole would be the person to make the decision. She had been a member of President Reagan's Cabinet for less than six months and had made a broad pledge at her Senate confirmation hearings to implement the administration's policies. And President Reagan's commitment to regulatory relief at NHTSA had been quite clear since the 1980 presidential election campaign.

There were early indications that the Reagan administration was in no hurry to resolve the issue. For instance, the acting administrator of NHTSA, Diane Steed, said, "I think that if the Court said anything to us, it was 'take your time.' "[4] Attorneys for State Farm included her statement in another legal brief to the D.C. Circuit requesting that the same panel (Judges Mikva, Bazelon, and Edwards) retain jurisdiction over the controversy. State Farm's brief questioned Steed's reading of Justice White's opinion while using her statement as ammunition for their charge that the administration was going to engage in delay tactics. Steed insisted that she was misquoted.[5]

Attorneys for State Farm also called attention to the comments of an anonymous "high-ranking aide" to President Reagan in an auto industry trade journal. The aide sought to calm the fears of auto industry officials by stating that "Mrs. Dole's marching orders on this one will be cut by the Vice President's Task Force on Regulatory Relief."[6] "The Administration plans to fight this challenge all the way," the aide said. "If we lose, the worst the car makers should expect might be an automatic seatbelt requirement some years down the road."[7]

The task force requested that OMB economists collect more evidence and better arguments to support deregulation.[8] Working under the supervision of David Stockman's aide Christopher De-Muth, the OMB group concluded that former Administrator Ray Peck lost the case by refusing to challenge the cost-effectiveness of air bags. This view was shared and publicized by the Pacific Legal Foundation, a conservative public interest group. Their managing attorney, Sam Kazman, editorialized in the *Wall Street Journal* that "had Mr. Peck openly questioned [NHTSA's] claim [about air bag effectiveness] as part of his basis for rescinding the rule, then it is highly unlikely that the Supreme Court would have ruled, as it did, that the agency had illegally failed to consider an air bag–only requirement."[9]

The potentially influential role of OMB in responding to the Supreme Court's remand was distressing to air bag advocates. At Senate hearings in September 1983, Senator John Danforth lamented that "OMB really makes the decisions on auto safety."[10] He complained further that "OMB has preempted decision making at the Department of Transportation."[11] To combat OMB influence, attorneys for State Farm demanded that Secretary Dole make available all past and future communications on passive restraints between OMB and the DOT.[12]

Dole chose not to respond directly to these charges. She instead used Diane Steed, who was nominated in the fall of 1983 by the White House to succeed Raymond Peck as the administrator of NHTSA. Steed (a former aide to Senator Robert Dole) was a career federal official who had served as chief of the regulatory policy branch of OMB during the Carter administration. She had come to NHTSA as acting administrator in early 1981 and helped delay the passive-restraint rule. Later she served as deputy administrator of NHTSA under former Administrator Peck and helped launch seat belt education programs. Steed proved to be an expert communicator as well as a policy maker. Her primary concern was process, and she helped protect Dole from political attack.

To outside observers, NHTSA appeared to be permeated with OMB influence. Steed and her special assistant, Erika Jones, were both ex-OMB employees, as was the new deputy administrator of NHTSA, Harold Smolkin. And Chris DeMuth of OMB seemed to be opposed to mandatory passive restraints, even though his testimony before Senator John Danforth was not absolutely clear on this point.

At hearings before Senator Danforth, Administrator Steed denied charges that OMB had usurped DOT's power and insisted that "there has been no predetermined outcome of this particular rule."[13] In response to a similar line of questioning, Chris DeMuth (OMB's director of Regulatory Affairs) acknowledged that OMB must approve new regulations but testified that the recommendation of an agency head or cabinet head normally "carries some weight."[14] Ralph Nader was skeptical and predicted that the future of passive restraints "will depend significantly on the ability of Transportation Secretary Elizabeth Dole to persuade [President Reagan] to drop his ideological opposition to federal safety standards."[15]

Roger Smith's Strategy

GM Chairman Roger Smith did not see the 1983 Supreme Court decision as the end of his company's campaign against mandatory passive restraints. Among consumer activists, Smith was considered "the chief obstacle to the widespread deployment of these automatic lifesavers."[16] Their fears were not misplaced: Smith was determined to provide Elizabeth Dole with a strong technical base to rescind the automatic restraint standard.

When Secretary Dole took control of the passive-restraint issue, General Motors Corporation sought to influence the course of her investigation. In a letter to Secretary Dole, Roger Smith called on DOT to "retain one, or preferably two, private consulting firms to conduct studies and report their conclusions by a date selected by you as to the air bag and alternatives." Smith was particularly concerned that Dole should be provided independent estimates of the costs and effectiveness of air bags relative to other restraint systems. GM was "prepared to turn over" to selected consulting firms "our data base, including all of our proprietary design, engineering, cost and field experience data."[17]

Leaders of the insurance industry perceived Smith's proposal to be another tactic in his long-term campaign against the air bag. In a letter disputing Smith's proposal, Vice President Donald Schaffer of Allstate Insurance Company warned that the consulting firm strategy would only lead to further delay. Schaffer took direct aim at Smith's reputation on safety:[18]

> *Mr. Smith is not a credible advisor on this subject. He is an avowed opponent of automatic crash protection. He cancelled the GM air bag program. He caused GM to submit the strategy [detachable automatic belts] upon which rescission of the safety standard was based. He included rescission of passive restraint requirements on his first "wish list" when this Administration took office. His suggestions should be ignored.*

Schaffer, along with other insurance industry representatives, urged Dole to reinstate the passive-restraint standard immediately. Although Dole did not take Schaffer's advice, she also rejected Smith's proposal in favor of her own process of decision making.

The "Safety Secretary"

Elizabeth Hanford Dole is an interesting person: born and reared in Salisbury, North Carolina, a political science major and May Queen at Duke University, a Harvard-trained lawyer, administrative assistant to Democratic Senator B. Everett Jordan of North Carolina, staff to President Lyndon Johnson's Committee on Consumer Interests, consumer adviser in the Nixon White House, and Commissioner of the Federal Trade Commission. In December 1975 the 39-year-old Elizabeth Hanford married the 52-year-old Senator Robert Dole, creating the so-called "Power Couple" that has drawn widespread media attention.

Elizabeth Dole is described as an "incredibly gracious and warm" personality who inspires a loyal following and has a knack for creating consensus where conflict is anticipated.[19] Some observers find it peculiar that a devout feminist and consumer advocate would assume a Cabinet position in the conservative Reagan administration. Few would disagree that Elizabeth Dole is a shrewd politician with a remarkable capacity to sustain her effectiveness in sharply different political and ideological environments.

In October 1983 Secretary Dole took personal control of the issue by committing DOT to a rule-making process with a specific timetable.[20] She called for written public comments and public hearings in three cities by the end of 1983. Three main alternatives were outlined in the October 1983 notice:

1. Reinstatement of the passive-restraint standard.
2. Amendment of the standard to ban permanently detachable seat belts or to require air bags.
3. Rescission of the standard again.

These main alternatives were accompanied by several additional options that could be pursued simultaneously with one of the three main alternatives. The additional options included:[21]

- A large-scale automatic restraint demonstration program.
- Federal legislation to require or encourage states to adopt mandatory seat belt use laws.
- A federal requirement that manufacturers offer automatic restraints as an option to car buyers.

The diversity of alternatives reflected Dole's determination to take a hard look at all possible courses of action—or at least her desire to persuade outsiders that alternatives were fairly considered.

To grapple with this sensitive issue, Dole formed a small team of eight policy advisers from DOT and NHTSA led by James Burnley, who had been promoted from chief counsel at DOT to the number two post in DOT (deputy secretary). The team included no career officials from NHTSA or DOT, and no disclosures of the team's deliberations were made, not even to other actors within the Reagan administration. According to one team member, "the consensus was that we would be subjected to outside pressures before we started if we let our discussions get out."[22] As a result, interest groups realized that their only avenue of communication to Dole's team would be through public hearings and written submissions.

Push for Belt Use Laws

The public testimony before Secretary Dole in December 1983 revealed widespread support for only one policy option: enactment of mandatory seat belt wearing laws. Auto industry officials advocated such laws as a substitute for passive restraints, while air bag advocates supported such laws as supplements to passive restraints. The key obstacle to their enactment was a lack of political enthusiasm for them among state legislators.

Representative John Dingell (D–Mich.) introduced a bill in Congress that would authorize categorical highway grants for states that adopted seat belt use laws.[23] The bill called for $15 million in grants in fiscal year 1985, $30 million in each of fiscal years 1986 through

1990, and $20 million in each of fiscal years 1991 through 1996. If states failed to enact seat belt use laws, the legislation proposed a gradual cutoff of funds. The grants would supplement the $10 million belt-promotion campaign that was already underway due to an earlier initiative by former NHTSA Administrator Peck. At Dole's hearings Congressman Dingell urged the secretary to support his approach. But Dingell's bill never passed the Congress, in part because it was superseded by Secretary Dole's final plan.

Dingell's bill was supported by each of the major automakers and the Motor Vehicle Manufacturers Association. Although auto industry officials publicly supported belt use laws, some observers were skeptical about the extent of the industry's enthusiasm. In late 1983 Michigan came very close to passing a seat belt use law but failed to do so. The *Automotive News,* a Detroit-based trade newspaper, opined that the Michigan bill would pass "if the American manufacturers [would apply] a little of their legislative muscle."[24]

Richard Peet, who had left the House staff to become president of Citizens for Highway Safety, was nonetheless optimistic that a sustained campaign for passage of seat belt wearing laws would work. An aborted attempt by Congress in 1973–1974 to offer incentive grants to states that enacted belt use laws was successful in inducing thirty state legislatures to consider such laws. Public furor over the interlock fiasco crippled the chances for passage of these bills and led to the withdrawal of incentive grants. Yet Peet believed that renewed interest in traffic safety in the 1980s made passage of belt use laws a distinct possibility. He also urged Secretary Dole to advocate belt use legislation.[25]

Ford Breaks Ranks

Most automakers reiterated their opposition to passive restraints at the Dole hearings. David Martin, director of automotive safety at GM, testified that the company's engineering resources were being transferred from passive restraints to interior padding schemes ("friendly interiors") that would protect both restrained and unrestrained drivers. Martin himself was attracted to "friendly interiors" as a less expensive way to provide the protection desired by air bag advocates. Martin said GM had concluded that the company's $160 million, fifteen-year investment in passive restraints had not produced devices that met with sufficient consumer acceptance.[26] Representatives from Chrysler and American Motors Corporation echoed GM's call for rescission of the passive-restraint requirements.

Ford Motor Company broke industry ranks at Dole's hearings by proposing a mandatory four-year demonstration of passive belts and air bags. As explained by Ford's new vice president, Helen Petrauskas, the Ford proposal called for all automakers to equip 5 percent of their new cars with passive systems. A special credit toward the 5 percent rule would be awarded to companies that chose the air bag. This proposal reflected Ford's renewed interest in air bags.[27]

Petrauskas argued that a large-scale, real-world testing program would answer any remaining technological questions and promote public understanding and acceptance of the new restraint systems. She thought a "technology-forcing" rule would be unproductive. The 5 percent proposal reflected, Petrauskas explained, "our struggle to find a way to work in an evolutionary way to high-volume air bag installation."[28] Evolution was viewed as necessary to build a supplier base and perfect the technology.

In addition to Ford's proposal, there was interest in air bags on Capitol Hill. Senators Danforth and Packwood steered an omnibus highway safety bill through the Senate Commerce Committee that required four large auto manufacturers (Ford, GM, Toyota, and Nissan) to offer air bags on some of their 1986 models. This bill was weaker than an earlier Danforth bill that required all 1986 models to have air bags. The revised bill resulted from compromise negotiations between Danforth and Senator Wendell Ford (D–Ky.), an opponent of mandatory air bags.[29] At Dole's hearings Danforth made clear that he supported some form of passive-restraint program, even though his compromise initiative was doomed by opposition from the Reagan White House.

Insurers Present Their Case

Insurers and air bag advocates recommended that Dole reinstate the suspended passive-restraint standard. Donald Schaffer of Allstate testified that the rule-making record provided Dole with only two "legally supportable" avenues: adoption of an air bag–only standard or reinstatement of the suspended standard.[30] Testifying on behalf of insurers, Yale economist Richard Nordhaus testified that "I can think of no act on the nation's agenda, short of keeping the peace, that would save more lives and prevent more disabling injuries more quickly than requiring automatic protection now."[31] Dr. William Haddon of the Insurance Institute for Highway Safety released results of a national public opinion survey showing that a majority of drivers believed that air bags or automatic seat belts should be made standard equipment. He urged reinstatement of the standard without delay.[32]

Insurers argued further that development work on automatic belts had been stymied by NHTSA's 1981 rescission order and the resulting regulatory uncertainty. GM had abandoned nondetachable and interlock-guarded belt designs such as those offered on 1978-1980 Chevettes in favor of new designs that could be detached easily and permanently. The company felt that these designs would be less annoying to motorists, though they would not achieve very high usage rates. GM officials made clear that their interests in detachable automatic belts were driven solely by regulatory compliance considerations. Without a passive-restraint standard, they would install manual belt systems in all their new cars.[33]

Insurers could, however, point to some highly promising automatic belt designs that were accumulating real-world experience. Toyota had already achieved considerable success marketing a motorized, nondetachable shoulder belt on the Cressida in the early 1980s. The design achieved high consumer ratings for comfort and convenience and achieved usage rates in excess of 90 percent. The major drawback of the system was its $350 cost, about a factor of three larger than the cost of detachable automatic belts.[34]

The pioneer in automatic belt development, Volkswagen, had sold 400,000 Rabbits with interlock-guarded automatic shoulder belts and knee bolsters from 1976 to 1981. Observational surveys revealed belt use rates for drivers and passengers of Rabbits in excess of 70 percent. Faced with increasing competition from small Japanese cars, VW lowered prices on all its 1982 models. As a result, the automatic belt system was downgraded from standard equipment to an $80 option. The sales rate dropped to about 5 percent of vehicles.[35]

A telephone survey of automatic belt owners commissioned by NHTSA found that the Volkswagen and Toyota systems received more favorable customer responses than did the GM Chevette system. After nine months of ownership, the percentage of owners who reported favorable attitudes toward their systems was 74 percent for the Toyota system, 77 percent for the VW system, and 49 percent for the Chevette system. The key complaint about the Chevette design was that the lap belt interfered with entry and exit from the car.[36]

A major limitation of automatic shoulder belts and knee pads is that many engineers believe that a lap/shoulder belt system offers better crash protection to occupants.[37] Despite this limitation, insurers could point to several automatic designs that were quite promising from the standpoint of consumer acceptance.

AMA Endorsement

For fifteen years sporadic efforts had been made, unsuccessfully, to persuade the American Medical Association to endorse mandatory passive restraints. After the 1983 Supreme Court ruling, an activist physician named Robert Vinetz renewed efforts to obtain an AMA endorsement.

Vinetz received only limited assistance from consumer and public health groups, who regarded his ambitions as unrealistic. He nonetheless pushed onward. Working with insurers, he assembled video tapes of air bag crashes, called AMA leaders throughout the nation, and distributed pro–air bag pamphlets to AMA delegates. According to Joan Claybrook, Vinetz was "the driving force in overcoming years of resistance" to the air bag within the medical community.[38]

Just as Elizabeth Dole was approaching a decision on the Supreme Court remand, the AMA voted to endorse the policy of mandatory passive restraints. This event was another small but important victory for the Coalition to Reduce Car Crash Injuries, a group of consumer activists and public health groups supported by insurers that ultimately joined Vinetz's campaign. The AMA endorsement further isolated the automakers as one of the few organized groups in society actively opposing mandatory passive restraints.

Mrs. Dole's Plan

Dole's team of policy advisers began their critical brainstorming sessions about policy options in January 1984. At the start of this process, "Mrs. Dole voiced no preconceived ideas about which alternatives she liked."[39] Although Dole was an active participant in these deliberations, the day-to-day leadership was provided by Jim Burnley. In an internal DOT legal analysis, the department made its position clear: Their reading of Justice White's opinion was that rescission was a hopeless strategy. Later, Burnley met privately with Chris DeMuth of OMB to explain the legal situation. After an intensive discussion, Demuth was still unconvinced.[40]

Demuth felt that Burnley's original legal argument was "ludicrous." The Supreme Court had only said that the decision must be made within certain parameters and be adequately explained. After a series of meetings, Burnley's legal position became more sophisticated. DeMuth recalls Burnley arguing that "given the rule-making record, which was being stacked by State Farm's attorneys, there was no viable way to justify a decision to rescind." After numerous discussions, it occurred to DeMuth that "Burnley seemed to be on a mission to prep me to convert a matter of policy discretion into a matter of law." At each meeting with DeMuth, Burnley "dribbled out" new arguments and evidence against rescission in what appeared to be a strategy to "wear me [DeMuth] out."[41]

By the spring of 1984, Dole's team of policy advisers was converging on a rather novel plan that called for reinstatement of the passive-restraint standard unless enough states passed compulsory belt use laws by 1989. There were two problems, however: (1) to issue such an unusual plan as a *final* rule without testing for public reactions, and (2) more time was necessary to lay the political groundwork for this plan within the Reagan administration.

In May 1984 Secretary Dole issued a supplemental notice of proposed rule making seeking comments on four new alternatives to reinstating the rule: (1) adoption of a mandatory demonstration program modeled after Ford's proposal; (2) a requirement that small cars be equipped with air bags on the driver side only; (3) an automatic restraint requirement with waivers for vehicles sold in states covered by mandatory seat belt use legislation; and (4) an automatic restraint requirement for all cars manufactured after a set date unless three-fourths of the states had enacted mandatory belt use laws prior to that date.[42]

Some career officials at NHTSA were disgruntled about these new proposals because they were not consulted. The matter was managed exclusively by Dole, Steed, Burnley and his staff.[43] Insurers complained that consideration of the new alternatives would further delay resolution of the issue beyond the July 11 pledge made by Secretary Dole.[44] Chris DeMuth of OMB made it clear to Burnley that he still favored rescission, but DeMuth had not yet figured out how to stop a determined Dole.[45]

After the supplemental notice was issued, the team needed some additional expertise to fashion the final rule. Dole personally recruited two career officials to assist in the effort: DOT attorney Neil Eisner was asked to help draft the rule and NHTSA analyst Barry Felrice was asked to help write the final regulatory impact analysis. Eisner and Felrice were informed of the emerging plan but, like other team members, were asked to discuss it with no one else until it was announced publicly.[46] Dole knew the risk of letting career officials in on the matter, but apparently felt that their expertise was worth the risk of leaks.

Elizabeth Dole took charge of the campaign to win support of her plan within the Reagan administration. Several weeks prior to the public announcement, Burnley met with OMB and White House staff in an effort of "neutralize" sources of potential opposition. During this critical period, OMB Director David Stockman was away on vacation. DeMuth at OMB felt he did not have the political clout to stop Dole, even though he disagreed with her plan and knew the President would have reservations.[47] Nevertheless, Burnley's discussions at OMB and the White House lasted for several days before Secretary Dole's scheduled meeting with President Reagan.

After an unexpected postponement of several days, Secretary Dole met with White House Chief of Staff James Baker and President Reagan. In preparation for the meetings, she took two precautions in the event of trouble. First, several demonstration cars with air bags and automatic belts were parked in the circle driveway at the White House. Second, an eminent physician with ties to the First Lady's family (Dr. Paul Meyer, Spinal Cord Center, Northwestern University Medical School) flew to Washington, was briefed on the issue, and was ready outside the Oval Office in case Dole would need a medical explanation of the need for passive restraints. Neither precaution proved to be necessary; Secretary Dole left the White House with President Reagan's endorsement of her plan.[48]

According to DeMuth, President Reagan was strongly opposed to Dole's decision but acquiesced only when Ed Meese told him that the Supreme Court had "compelled" such a decision. DeMuth explains that James Baker and Ed Meese "let Mrs. Dole have her way because it would have been politically too controversial to fight her."[49] David Stockman, who was on vacation at the time, never became deeply involved in the matter.

On July 11 Secretary Dole announced a new final rule requiring passive restraints to be installed in 10 percent of 1987 models, 25 percent of 1988 models, 40 percent of 1989 models, and 100 percent of 1990 and later models. During the phase-in period, the rule encouraged manufacturers to comply with air bags or other "friendly interiors" (instead of automatic belts) by counting each car with the nonbelt technologies as 1.5 units.[50]

Dole's plan also included a rescission provision designed to encourage states to adopt mandatory seat belt use laws. If states representing two-thirds of the nation's population enacted mandatory seat belt use laws before April 1, 1989, the requirement for automatic protection would be removed. State laws counting toward the two-thirds threshold were expected to satisfy six criteria:

1. No waivers except for medical reasons.
2. An effective date no later than September 1, 1989.
3. A minimum penalty of $25 for noncompliance.

4. A civil litigation provision allowing for mitigation of any damages sought by injured yet unbelted plaintiffs.
5. A program to encourage compliance with the law.
6. An evaluation program to determine the law's effectiveness.

Secretary Dole also announced that $40 million per year in DOT and private funds would be made available for a nationwide cooperative educational campaign to encourage safety belt use and mandatory usage laws.[51]

The decision by Elizabeth Dole reflected both a deep commitment to safety and very astute politics. She offered something to everyone: Insurers and air bag advocates were awarded a passive-restraint standard while the auto companies and belt use advocates were awarded a national belt use promotion campaign. The design of the plan assured safety progress: Either passive restraints were coming or mandatory belt use laws were coming. And the plan was constructed so that both policies might be in place simultaneously, at least until April 1989—after the 1988 elections.

A Tactical Slipup

The Dole team planned several dozen phone calls to key stakeholders the morning prior to her public announcement. Such calls are a key sign of professionalism in Washington politics, as they are a gesture of respect for other powerbrokers and permit outsiders to prepare their reactions for the press.

A major oversight occurred when the Dole team neglected to contact the chairmen in the House and Senate appropriations committees (William Lehman of Florida and Mark Andrews of North Dakota) prior to the press announcement. If Secretary Dole had simply planned to reissue the passive-restraint rule, the calls she made to Danforth in the Senate and Dingell and Wirth in the House might have been sufficient. A key plank in the Dole plan was, however, a $40 million seat belt education program, including $20 million from the auto industry and $20 million from "the public sector." Although the leaders of the auto industry were contacted (and were apparently quite surprised to learn about this $20 million contribution!), Lehman and Andrews were inadvertently ignored (even though they held the key to the purses of "the public sector").[52]

When Dole's advisers belatedly contacted Lehman and Andrews, it was too little, too late. After extensive communications throughout the summer about the fiscal year 1985 budget, it became apparent that Secretary Dole could not win congressional support for her $20 million request. Attempts by Secretary Dole's advisers to run around Lehman and Andrews only led to more bitter resistance.[53]

According to one of Dole's advisers, they failed to get the $20 million for several reasons. First, Lehman and Andrews were understandably offended by not being consulted prior to Dole's announcement. Second, key congressional staff feared that these funds might be used in an effort to kill the air bag by promoting mandatory belt use legislation. Third, Lehman, a Democrat, was reluctant to appropriate $20 million for public relations activities in an election year, especially since Mrs. Dole might prove to be quite an asset to many Republican candidates in the House and Senate. Finally, there was a serious question about how this amount of money could be usefully spent, and what the measures of effectiveness might be.[54]

In the final analysis, Congress appropriated only $7.5 million for NHTSA's seat belt education program. Even this amount represented only $2.5 million in new monies, considering former Administrator Peck had retained a commitment of $5 million for belt use education in FY 1984 (despite congressional skepticism about his program). Going into the fall of 1984, it was beginning to appear that the auto industry would have to be the principle source of resources in the campaign to enact mandatory belt use laws.

Detroit's Strategy Shifts to the States

Secretary Dole's decision left the auto manufacturers with only three practical options for continuing their opposition to mandatory passive restraints. They could fight Dole's plan in federal court, they could fight the plan in Congress, or they could initiate a lobbying campaign on behalf of mandatory seat belt wearing laws. Secretary Dole had invited the industry to choose the latter course by announcing (unilaterally) that the "private sector" would contribute $20 million to her safety belt promotion campaign. Almost overnight, Detroit's lukewarm support for seat belt legislation was transformed into an aggressive, $15 million lobbying campaign throughout the nation. A new organization, Traffic Safety Now, was created with industry support to distribute grants to grass-roots, prosafety groups throughout the nation.

Traffic Safety Now is a unique organization whose purpose is to reduce deaths and injuries by promoting mandatory belt use laws. The organization was formed by a consortium of domestic and foreign carmakers that made contributions in proportion to market share.[55] It was originally run by ex-auto executives and had close ties with James Johnston, GM's new vice president for governmental affairs and an ex-lobbyist from Washington with extensive experience in international affairs. He was apparently brought to GM by chairman Roger Smith to help improve the company's image in Washington for social responsibility.

The stepped-up campaign for belt use laws came at an opportune time. State legislators were already sensitized to the traffic safety problem as a result of debates about child-restraint laws, drunk-driving crackdowns, and minimum drinking age legislation. Former administrator Peck's belt promotion campaign, which Secretary Dole proposed to enlarge, had already reinforced public awareness of the safety issue. Before Dole's decision was announced, New York had become the first state to adopt a mandatory belt use law.

Insurers and air bag advocates supported Dole's passive-restraint mandate but opposed what they called the "trapdoor": the automatic rescission provision.[56] Arnold and Porter, representing State Farm, sued DOT in the D.C. Circuit seeking a judicial order against the trapdoor. The state of New York also sued DOT for permitting manufacturers to install detachable automatic belts. They argued that it was "arbitrary and capricious" of Secretary Dole not to either require air bags or disallow detachable automatic belts.

Originally, Claybrook and State Farm sent signals that they would oppose Dole's campaign to enact seat belt wearing laws. Because this stance would have been very difficult to defend credibly, they changed their position to favor belt use laws but urged state legislators to pass weak laws that would not satisfy the criteria in Secretary Dole's plan. Even this position seemed a bit hypocritical in some circles.

Fear of the "Trapdoor"

New York State passed a mandatory belt use law—the first in the nation—just before Secretary Dole revealed her final plan about passive restraints. Since the final plan called for repeal of the passive-restraint rule if two-thirds of the U.S. population was covered with laws within five years, the New York law was potentially significant. As the nation's second most populous state (17.7 million residents or about 7 percent of the U.S. total), New York would make a substantial contribution to the two-thirds threshold. Attorneys for State Farm calculated that the standard would be rescinded if only the 16 most populous states became covered by belt use legislation. The trapdoor imagery was used to dramatize this scenario.[57]

To help make sure that New York's population could be counted toward the two-thirds threshold, Secretary Dole included a special waiver provision in her final plan. It stated that DOT "would consider granting a waiver from the minimum requirements for any state that, before August 2, 1984, has passed a mandatory use law that substantially complies with those requirements." This waiver opportunity was important because the New York law contained a maximum fine of $50 while DOT's criteria called for a minimum fine of $25.[58]

After Dole's decision New York Governor Mario Cuomo issued an executive memorandum of July 12, 1984, that approved the state's belt use law. Cuomo stated that "the legislative intent of the bill is to require use of safety belts across the lap. Thus, if the shoulder harness causes discomfort, it could be placed behind the person."[59]

On the basis of this memorandum, the Insurance Institute for Highway Safety petitioned Secretary Dole to disqualify New York's law from counting toward the two-thirds threshold. The Institute argued that motorists would have substantially less crash protection without the shoulder belt and that the law would be much more difficult to enforce with only a lap belt requirement. Institute President Dr. William Haddon charged that a DOT decision to approve New York's law would "make a mockery of the minimum requirements specified in the standard."[60]

Later in 1984 Governor Cuomo wrote Secretary Dole urging her to reconsider the final rule on passive restraints. Cuomo expressed concern that the rule would pit supporters of mandatory seat belt use against advocates of automatic protection. He recommended that Dole encourage "the accelerated development and implementation of both active and passive restraint systems in all automobiles to give the greatest protection possible to the largest possible number of people."[61]

Cuomo's concern was not theoretical. By January 1985 two more populous states, New Jersey and Illinois, enacted mandatory belt use legislation. In New Jersey the law was amended to require only a $20 fine in an effort to prevent DOT from counting the state's population toward the two-thirds rescission threshold. The New Jersey law was also written to permit only "secondary" police enforcement (i.e., police were permitted to issue fines for nonbelt use only if the motorist was already being cited for some other traffic violation). New Jersey's Republican governor, Thomas Kean, made clear that if a choice must be made, he preferred a federal rule requiring air bags to a belt use law, which he regarded as difficult to enforce.[62]

Many air bag advocates perceived that a secondary enforcement provision would weaken a law enough to preclude its counting toward the two-thirds threshold. In fact, the criteria in the Dole plans were silent on secondary enforcement. While it may have been difficult politically to count such laws, there was nothing in the rule that would have precluded such an action.

·The provisions of the Illinois law also seemed to be at odds with the minimum requirements in Dole's passive-restraint rule. The law contained a maximum $25 fine, precluded reduction of damage awards in civil suits based on nonbelt use, and permitted only secondary police enforcement. Since Illinois is a very populous state (11.4 million residents or about 5 percent of the U.S. total), air bag advocates were eager to learn whether Dole would approve the Illinois law.

Secretary Dole did not respond publicly to criticism of the "trapdoor" because the provision was the target of litigation in the D.C. Circuit. At congressional hearings, NHTSA Administrator Diane Steed told legislators that there was ample time available before April 1, 1989, to evaluate laws against the criteria and make recommendations to states whose laws did not meet the criteria.[63] Dole and Steed refused to make immediate decisions on whether particular state laws met the minimum criteria specified in the final rule. Steed did make available to each state a model seat belt use law that was written to satisfy the criteria specified in Dole's plan.

The refusal of Secretary Dole to answer questions about the adequacy of particular state laws was a source of much frustration among air bag advocates. For example, former NHTSA Administrator Joan Claybrook left the agency in 1981 and later assumed the presidency of the public interest group, Public Citizen. In this capacity she engaged herself in active lobbying in several states to ensure that the belt use laws that were passed would not satisfy DOT's minimum requirements. In testimony before the U.S. Congress, she urged senators and congressmen to pressure Secretary Dole to rule on the acceptability of particular state laws. Without a formal ruling, Claybrook feared that Dole might later waive or change the criteria in order to count more states.[64]

Senator Danforth was also concerned that the automobile manufacturers would persuade enough states to pass seat belt laws in order to cause repeal of the passive-restraint rule. He reported that General Motors Corporation was using the Saturn plant and the potential economic stimulus to the host state as a "tool" to get states to pass belt use legislation. The Missouri Senate, according to Danforth, passed a weak belt use bill in 1985 in an "effort to satisfy General Motors, in order to please them in this frenzied effort to try to get the Saturn plant."[65] During the 1985 legislative sessions, bills to require safety belt use were introduced for consideration in twenty-nine states.[66]

The Lehman-Dole Feud Intensifies

A key ingredient of Secretary Dole's campaign for mandatory seat belt use legislation was congressional funding of the National Occupant Protection Program. This program included three components: educational and informational campaigns directed at states and communities, with emphasis on building coalitions of elected officials, business leaders, and volunteer organizations; networking with national organizations such as the National Red Cross and the American Public Health Association; and renewed media messages tailored to the educational process at state and community levels.[67] To implement this program, Dole continued to ask Congress for $20 million in fiscal year 1985 and $20 million in fiscal year 1986. She initiated a "full court press" on the Hill, despite previous denials of this request.[68]

William Lehman (D–Fla.) and Mark Andrews (R–N.Dak.) remained skeptical about the purposes and ultimate effectiveness of Dole's educational program.[69] Former NHTSA Administrator Peck had set a goal of 25 percent belt use as a justification of his $30 million educational campaign during fiscal years 1984–1986. The agency's data showed only a slight increase in belt use, from 14 to 15 percent, after eighteen months of the campaign.[70] Moreover, Lehman and Andrews remained concerned that the federal monies requested by Dole would be transferred to groups that would then lobby for enactment of belt use laws, even though Administrator Steed denied that the funds would be used directly for lobbying. It was also not clear that any of the funds would be devoted to education about automatic restraint systems.

In testimony before the House Appropriations Committee, Joan Claybrook attacked Dole's $40 million proposal. She said the request was grossly inflated and that many of the proposed activities would occur anyway without federal support. If the program were funded, Claybrook urged the Appropriations Committees to specify that specific amounts of money be spent for education about air bags. She charged that DOT had ignored the direct instructions of appropriations committees to spend an equal amount of money in fiscal year 1985 on seat belt use and passive restraints.[71]

Claybrook's opposition to belt use education irritated many industry officials. According to Roger Maugh of Ford: "She never placed a high priority on reducing fatalities or injuries due to lack of belt use."[72] Maugh argued that belts were critical because they positioned the occupant for air bag deployment and offered protection in nonfrontal crashes where air bags are not designed to inflate. In contrast, Claybrook saw belt use education as a wasteful investment (because it would not increase belt use) and as a diversion from passive protection.

Congressman Lehman attempted to use Secretary Dole's funding requests as a lever to induce Dole to announce which state laws—if any—satisfied the minimum requirements of her passive-restraint standard. In an April 3 letter to Dole, Lehman intimated that his committee would not release $7.5 million in previously appropriated funds until she answered his questions about state seat belt laws.[73]

Lehman's letter irritated John Dingell (D–Mich.), the House's arch foe of mandatory passive restraints. Dingell sent his own letter to Dole stating that the Appropriations Committee had no legal power to place such strings on money for DOT's belt use campaign. Dingell urged Dole not to respond to Lehman's letter.[74]

The issue became increasingly delicate as more and more states passed belt use laws. By August 1985, fifteen states had passed mandatory seat belt bills of some sort. Despite the pressures from Claybrook, Lehman, and attorneys for State Farm, Secretary Dole's attorneys made clear that she did not intend to rule on the adequacy of particular state laws until after the D.C. Circuit decided legal challenges to her final rule. Originally Dole had said that the department would review each state law as it was passed to determine whether it met the minimum criteria established by the regulation. (In the rush to get the rule out by the July 11th deadline, the rule had said one thing and the preamble to the rule said another.) But given the litigation and the growing interest in belt use legislation, Dole decided to wait.

Some critics argue that Dole departed from her neutral position vis-à-vis air bags and belt laws by refusing to rule on the adequacy of state laws.[75] Dole's supporters counter that such rulings would have undercut Detroit's incentive to enact belt use laws. One of Dole's advisers believes that "deep down, Mrs. Dole wanted both air bags and mandatory seat belt wearing laws."[76] The irony is that both sides would ultimately charge that Dole was not being neutral in her implementation of the rule.

Lehman made Dole pay a price for her silence. Originally a 1985 supplemental appropriations bill contained a provision that would have required DOT to rule on whether enacted mandatory belt use laws met DOT criteria. Dole and Dingell made sure that this provision was ultimately removed. Instead the bill rescinded $7.5 million from Dole's seat belt education program, as Lehman had threatened.[77]

Surprise in Sacramento

As the trend to seat belt legislation gained momentum, the legislative outcome in California—whose population accounts for 10 percent of the U.S. total—became increasingly important. Traffic Safety Now and its industrial supporters were pushing for a strong law that Dole could count toward the two-thirds threshold without question. Air bag advocates saw California as a critical test of their ability to save the passive-restraint requirements. Andrew McGuire, an injury prevention advocate, organized the California Coalition to Reduce Car Crash Injuries to lobby California legislators. He was assisted by the active efforts of Joan Claybrook and Ralph Nader.

Dr. William Haddon, Jr., also wrote an editorial for the *Los Angeles Times* of October 31, 1984, five months prior to his untimely death. Entitled "A California Seat Belt Law," the editorial warned the citizens and legislators of California that their decision to enact a seat belt use law could result in a nationwide setback for highway safety. He emphasized that California's law could activate the "trapdoor," thereby depriving all U.S. citizens of automatic crash protection. "The best solution," Haddon argued, "is to provide automatic protection, including air bags, as baseline protection for everyone, with seat belts as a supplement for those who will use them."[78]

Haddon's indifference to seat belt laws was appalling to safety officials in Detroit. According to Roger Maugh of Ford: "Haddon was more worried about getting air bags into cars than saving lives. He suffered from the same myopia as Claybrook."[79] Many industry observers were bewildered that so-called "safety advocates" could actually lobby against seat belt laws. They believed safety belts should be thought of as baseline protection for everyone and air bags as a supplemental protection for those willing to pay the price.[80]

The campaign of air bag advocates in California achieved results. A provision was included in the California seat belt law that called for automatic rescission of the law if DOT attempted to count California's population toward the two-thirds threshold for purposes of rescinding the passive-restraint requirements. (The idea for such a "reverse trapdoor" was actually borrowed from antiregulation lobbyists in Michigan and Missouri, who had succeeded in adding provisions calling for repeal of belt use laws if DOT's rule was not rescinded.) Another provision, one that in all probability will be struck down by the courts due to federal preemption, called for all new cars sold in California to be equipped with air bags.[81]

A key to the legislative battle in California was the position of Assembly Speaker Willie Brown. As sponsor of what became the "seatbelt/passive restraint" bill, Brown sided with air bag advocates because he wanted both a belt use law and automatic restraints. He took the case to his colleagues, and the issue was finally resolved late in the 1985 legislative session. The bill ultimately passed just hours before the close of the session by a vote of 48 to 21 in the California Assembly and 24 to 10 in the California Senate. When Republican Governor Deukmejian indicated he would sign the bill, a crucial step had been taken toward dismantling the "trapdoor."[82]

California legislators made clear that they preferred their citizens to be protected by both air bags and mandatory seat belt use. The impact of this message was powerful. After California's law was enacted, attorneys for Secretary Dole made clear during oral argument before the D.C. Circuit that DOT would not attempt to count California's population toward the two-thirds threshold.

Roger Maugh of Ford commented:[83]

> *Traffic Safety Now and its industrial supporters could have killed the [California] bill. We sat down and talked about it. We concluded it would be a plus for safety, despite the crazy add-on provisions. Hence we supported the overall bill.*

The philosophy of TSN was to support belt use legislation regardless of what form it took. The idea was to get the best bill possible and circle back later with corrective amendments.[84] That was the lesson drawn from earlier efforts to pass child-restraint legislation.[85] The outcome in California was the event that finally persuaded everyone that the standard was not going to be rescinded.[86]

A Go-Ahead from a Different D.C. Circuit

The original legal challenges to Secretary Dole's decision of July 1984 were made by insurers and the state of New York. The new three-judge panel of the D.C. Circuit formed to hear the case included Judge Abner Mikva, who had written the earlier opinion overturning Peck's decision, and Judges Edward Tamm and Kenneth Starr. Tamm was a conservative veteran of the D.C. Circuit and Starr a recent appointee of President Reagan's.

Oral argument was heard by the panel in early 1985, and a decision was expected by summer 1985. Judge Tamm died unexpectedly before the panel's opinion was finished, and the D.C. Circuit was therefore compelled to appoint a new judge to the panel and call for a second round of oral argument. (If the original panel's decision had been unanimous, it is likely that Starr and Mikva would have released an opinion for the original panel. The formation of a second panel was a signal that the original panel was not in agreement.)

The new judge appointed to the panel was Antonin Scalia, a conservative appointee of President Reagan's and a former law professor at the University of Chicago Law School. Scalia participated in the second round of oral argument in early 1986. The speculation of many attorneys in the case was that the conservative coalition of Starr and Scalia was likely to uphold Dole's decision, forcing Mikva to write a dissenting opinion. A decision was expected in the summer of 1986.

That summer Chief Justice Warren Burger resigned unexpectedly from the Supreme Court. President Reagan nominated Justice Rehnquist to become Chief Justice and Judge Scalia of the D.C. Circuit to take Rehnquist's seat on the Court. These events occurred before the D.C. Circuit panel had published its opinion on passive restraints. The panel did ultimately deliver an opinion, but not until September 1986.

In a surprising holding, a unanimous panel concluded that most of the legal challenges to the "trapdoor" in Dole's plan were not "ripe" for judicial review.[87] Judge Starr's opinion for the Court stated that no parties would be seriously harmed if the court refused to review the automatic rescission provision until it was actually employed by DOT. The opinion stressed that it was very unlikely that rescission would occur given that most state belt use laws did not appear to satisfy DOT's minimum criteria. A footnote in the opinion warned DOT that any steps to alter the criteria would be tantamount to changing the rules in the middle of the game.

This "ripeness" ruling proved to be a victory in unexpected form for insurers and air bag advocates. Though State Farm's attorneys had urged the court to invalidate the "trapdoor," the panel did not do so but instead sent an implicit signal that weak laws would not be a satisfactory basis for rescinding the passive-restraint standard.

A majority of the panel (Scalia and Starr) also upheld Secretary Dole's decision to allow detachable automatic belts to be installed by manufacturers under the standard. Although Judge Mikva dissented on this point, the U.S. Supreme Court refused to hear an appeal from New York State. In effect, the D.C. Circuit gave Dole's plan the final go-ahead.

Conclusion

Secretary Dole's decision on occupant-restraint regulation was scrutinized, in one way or another, in virtually every conceivable power center: the Office of Management and Budget, the Reagan White House, the halls of the U.S. Congress, the federal courts, and state legislatures. Unlike previous attempts to resolve this issue over the past twenty years, the Dole plan withstood the test of time.

Dole's success reflects not just her personal charm but her shrewd approach to regulatory politics. She saw that any plan she proposed could be vulnerable at numerous decision points. Her key insight was to reduce her vulnerability by offering each powerful interest group something they desired. Though she did not achieve a political consensus, she was able to win some converts while softening the opposition of her remaining critics. Once her plan was enacted, she engaged in tough political tactics to buy precious time for implementation.

Elizabeth Dole's plan achieved for America what most developed nations had done ten years earlier: widespread enactment of mandatory belt use laws for adults. Unlike her predecessors, she refused to accept the conventional wisdom that such laws were politically infeasible. In effect, she demanded that Detroit put its checkbook and immense lobbying power behind its rhetoric in favor of belt use laws. Although she committed a serious tactical error in her relations with the appropriations committees of Congress, it is not clear how much that error impeded the campaign for belt use legislation.

Mandatory seat belt wearing laws were not the sole accomplishment flowing from Dole's plan. As we shall see in the next chapter, an amendment to her plan facilitated a breakthrough in Detroit's posture toward the air bag. Once again, the amendment occurs through a classic exercise in pluralistic politics.

Endnotes

1. "Rescission of Air Bag Rule Is Hit," *Washington Post* (June 25, 1983), p. A1.
2. Ibid.
3. "Despite Supreme Court Ruling, Airbag Fight Will Continue," p. 45.
4. "New Issue Emerges in 'Passives' Ruling," *Status Report: Highway Loss Reduction* (August 16, 1983), pp. 1, 7.
5. Ibid., pp. 7–8.
6. "Despite Supreme Court Ruling, Airbag Fight Will Continue," p. 45.
7. Ibid.
8. Ibid.; "Time to Act on Airbags," *New York Times* (August 29, 1983), p. A18.
9. Sam Kazman, "Deflating the Claims of the Air-Bag Studies," *Wall Street Journal* (July 21, 1983), p. 22.
10. "Danforth Assails Roadblocks to Air Bags," *Status Report: Highway Loss Reduction* (September 20, 1983), p. 2.
11. Ibid., p. 2.
12. Letter from James F. Fitzpatrick, of Arnold and Porter, to Secretary Elizabeth Dole, August 16, 1983.
13. "Danforth Assails Roadblocks to Air Bags," p. 2.
14. Ibid.
15. Ralph Nader, "Air Bags," *Washington Post* (July 8, 1983).
16. Ralph Nader and William Taylor, *The Big Boys: Power and Position in American Business* (New York: Pantheon Books, 1986), p. 138.
17. "Offering a Lift," *Wall Street Journal* (September 12, 1983), p. 24.
18. Letter from Donald Schaffer, vice president, Allstate Insurance Company, to Secretary Elizabeth Dole, 1983.
19. Philip Weiss, "Charming Her Way to the White House," *Washington Monthly* (September 1987), p. 43.
20. *Federal Register*, 48:48622–48641 (1983).
21. Ibid.
22. Interview of Phil Haseltine, July 21, 1988.
23. "Dingell Bill Promotes Mandatory Seat Belt Laws," *Status Report: Highway Loss Reduction* (November 9, 1983), p. 5.
24. "Detroit and Seat Belts," *Automotive News* (September 25, 1983).
25. Testimony of Richard C. Peet, president, Citizens for Highway Safety, DOT Hearings on Occupant Crash Protection, December 7, 1983.
26. Remarks of David E. Martin, General Motors Corporation, DOT Hearings on Occupant Crash Protection, December 5, 1983.
27. "Ford Proposes Four-Year Test of Two Million Cars Equipped with Passive-Restraint System," *Automotive News* (December 5, 1983), pp. 1, 58.
28. Interview of Helen Petrauskas, September 3, 1987.
29. "Air Bags," *Congressional Quarterly Almanac* (1983, 1984), p. 544; "Senate Panel Advances Air Bag Proposal," *Status Report: Highway Loss Reduction* (October 4, 1983), p. 7.
30. "Auto Insurers Advise DOT Officials to Stay with Current Rule," *Status Report: Highway Loss Reduction* (December 19, 1983), p. 1.
31. Statement of Professor William Nordhaus, Yale University, DOT Hearings on Occupant Crash Protection, December 6, 1983.

32. Statement of William Haddon, Jr., president, Insurance Institute for Highway Safety, DOT Hearings on Occupant Crash Protection, December 5, 1983.
33. Interview of David Martin, September 3, 1987.
34. *Automatic Safety Belt Usage in 1981 Toyotas*, U.S. Dept. of Transportation, NHTSA, February 1982, DOT-HS-806-146.
35. Statement of Volkswagen of America, Inc., Public Hearing on Occupant Crash Protection, U.S. Dept. of Transportation, NHTSA, December 6, 1983.
36. *Automatic Safety Belt Systems: Owner Usage and Attitudes in GM Chevettes and VW Rabbits*, U.S. Dept. of Transportation, NHTSA, May 1980, DOT-HS-805-399.
37. Ibid.
38. Speech by Joan Claybrook, "Stretching the Physician's Role: Influencing Health Policy by Giving Priority to Injury Prevention," Stanford University, January 24, 1986.
39. Interview of Phil Haseltine, July 21, 1988.
40. Interview of Christopher DeMuth, July 28, 1988.
41. Ibid.
42. *Federal Register*, 49:20460–20470 (1984).
43. Interview of Barry Felrice, July 23, 1987.
44. "Administration Is Criticized for Delay on Automatic Restraints," *Status Report: Highway Loss Reduction* (May 12, 1984), pp. 2–3.
45. Interview of Christopher DeMuth, July 28, 1988.
46. Interview of Phil Haseltine, July 21, 1988.
47. Ibid. Interview of Christopher DeMuth, July 28, 1988.
48. Ibid. Interview of Phil Haseltine, July 21, 1988.
49. Interview with Christopher DeMuth, July 28, 1988.
50. *Federal Register* 49:28,962 (1984).
51. Press Release, "Dole Calls for Occupant Protection in Cars Beginning in 1986; Urges State Seat Belt Laws," Office of Public Affairs, U.S. Dept. of Transportation, July 11, 1984.
52. Interview of Charles Hurley, June 22, 1988.
53. Ibid.
54. Interview of Phil Haseltine, July 21, 1988.
55. Interview of Charles Spillman, June 30, 1987.
56. William Haddon, Jr., "The Trapdoor," *Baltimore Sun*, reprinted in *Status Report: Highway Loss Reduction* (January 19, 1985), p. 2.
57. Petitioners' Principal Brief (Arnold and Porter), *State Farm* v. *Elizabeth Dole* (D.C. Circuit, 1984), p. 28, Figure 1.
58. "Petition Would Deny Waiver for New York's Seat Belt Use Law," *Status Report: Highway Loss Reduction* (November 3, 1984), pp. 1–2.
59. Ibid.
60. Ibid.
61. "Cuomo Urges Dole to Reconsider Restraint Ruling," *Status Report: Highway Loss Reduction* (November 24, 1984), p. 7.
62. "New Jersey Becomes Second State to Enact Seat Belt Use Law," *Status Report: Highway Loss Reduction* (March 24, 1984), p. 3.
63. Statement of NHTSA Administrator Diane Steed, in *NHTSA Authorization*

and Means of Improving Highway Safety, Hearings before Senate Committee on Commerce, Science and Transportation, 99th Congress, First Session, February, 21, 1985, pp. 3–25.

64. Statement of Joan Claybrook, president, Public Citizen, ibid., pp. 54–56.
65. Statement of Senator John Danforth, ibid., pp. 5–6; also see "Senator Critical of Seat Belt Use Laws That Sacrifice Air Bags," *Status Report: Highway Loss Reduction* (March 16, 1985), pp. 4–5.
66. Testimony of NHTSA Administrator, Diane Steed, *DOT and Related Agencies Appropriations for 1986*, Hearings before House Committee on Appropriations, 99th Congress, First Session, 1985, p. 34.
67. Ibid., pp. 34–5.
68. "DOT and NHTSA Appropriations," *Automotive Engineering* (May 1985), p. 27.
69. "DOT's Seat Belt Program Sharply Attacked in House, Senate Hearings," *Status Report: Highway Loss Reduction* (April 6, 1985), pp. 1–4.
70. Ibid.
71. Testimony of Joan Claybrook, *NHTSA Authorization and Means of Improving Highway Safety*, pp. 56–57.
72. Interview of Roger Maugh, July 8, 1987.
73. Letter from Congressman William Lehman to Secretary Elizabeth Dole, April 3, 1985.
74. "Statehouses Pumped Up for Battle on Air Bags," *National Journal* (April 27, 1985), p. 916.
75. Interview of Charles Hurley, June 22, 1988.
76. Interview of Phil Haseltine, July 21, 1988.
77. "$13.02 Billion Fiscal '85 Supplemental Cleared," *Congressional Quarterly Almanac* 1985, 1986, p. 359.
78. William Haddon, Jr., "A California Seat Belt Law," *Los Angeles Times* (October 31, 1984).
79. Interview of Roger Maugh, July 8, 1987.
80. Interview of Don McGuigan, September 3, 1987; interview of David Martin, September 3, 1987.
81. Assembly Bill Number 27, Proposed Conference Report Number 1, September 10, 1985, California Legislature, 1985–86 Regular Session.
82. Stephen P. Teret, "Victory in California," *Injury Prevention Network* (Summer/Fall 1985), p. 1.
83. Interview of Roger Maugh, July 8, 1987.
84. Interview of Helen Petrauskas, Chairman of the Board, Traffic Safety Now, September 3, 1987.
85. Statement of NHTSA Administrator Diane Steed, *NHTSA Authorization and Means of Improving Highway Safety*, p. 5.
86. Interview of Phil Haseltine, July 21, 1988.
87. *State Farm* v. *Dole*, 802 F.2d. (D.C. Circuit, 1986), p. 474.

Chapter 9

A NEW STRATEGY AT FORD

Modern American pluralism is a mixture of government and private capitalism. While our political system is designed to hinder radical reform, our economic system—governed as it is by the profit motive and the pressures of competition—is quite capable of radical change. Because a relatively small number of business leaders often control huge amounts of resources, competition can cause swift and fundamental changes in investment strategies.

The revival of the air bag in the 1980s illustrates the powerful influence of new corporate strategies. A new generation of leadership at Ford Motor Company decided in the early 1980s that it wanted to disarm its critics in Washington and provide more value to the customer by offering improved driver-side air bags as standard equipment on all its lines. Accordingly, Ford persuaded NHTSA to modify its rule to facilitate their new policy during the 1987–1994 period. Ford's new corporate policy then triggered competitive pressure throughout the industry, and the spirited resistance to air bags at Chrysler and General Motors quickly dissolved. In a few short years the troubled air bag industry was saved from near bankruptcy and transformed into a flourishing multibillion-dollar industry.

Suppliers Without Customers

When the Reagan administration took office in January 1981, the air bag industry was on the verge of extinction. It was impossible for a consumer to buy a new car with air bags from any domestic or foreign manufacturer. General Motors had terminated its air bag development program and was working exclusively on automatic belts. Chrysler remained opposed to air bags, as it had throughout the 1970s. Although Ford had retained a small air bag research effort, it was also looking at automatic belts as the primary means of compliance with NHTSA's passive-restraint requirements. And even the future of automatic belts was in doubt as the Reagan administration moved rapidly to delay and then rescind the passive-restraint requirements.

These decisions were devastating to air bag suppliers. Eaton Corporation and Allied Chemical Corporation dropped out of the air bag business in 1979. The surviving suppliers were small and financially insecure. As the air bag industry teetered on the brink of collapse, career officials at NHTSA scrambled to keep the industry alive. With the support of NHTSA Administrator Raymond Peck, agency officials looked for methods to funnel development funds to creative air bag suppliers.

The key air bag supporter at NHTSA was Michael Finkelstein, a career public servant who ran rule making under Claybrook and R&D under Peck. Finkelstein worked on three programs to keep the air bag technology alive: a retrofit program for state police cars, a research program to develop air bags with mechanical crash sensors, and a government-funded demonstration program for air bags in new government cars.

In November of 1982 Administrator Peck announced plans to install driver-side air bags in several state police fleets. NHTSA subsequently awarded a contract of $1 million to Romeo-Kojyo Company of Tempe, Arizona, for 500 air bag retrofit kits.[1] The retrofit systems were successfully tested in 35 mph car and sled tests. By early 1984 Romeo-Kojyo Company was offering the retrofit system to state police departments.[2]

NHTSA officials were also intrigued with a new mechanical crash sensor for air bags developed by the Breed Corporation, a small New Jersey engineering firm. The new design substituted mechanical crash sensors for more complex electronic components, thereby cutting drastically the costs of mass-producing air bag systems. In early 1984 NHTSA awarded Breed a $588,000 contract to further develop this new air bag concept. The system was to be crash tested and ultimately installed in state police cars. When a series of crash tests and rough road driving proved successful, Finkelstein announced that NHTSA would equip a police fleet of 100 Chevrolet Impalas with the Breed System. Expansion to a fleet of 1,000 cars was mentioned as a possibility.[3]

Peck and Finkelstein also created a NHTSA-funded program whereby a fleet of 5,000 cars purchased by the federal government through the GSA would be equipped with air bags.[4] This was a good idea in theory, but it would have accomplished nothing without a car manufacturer to build cars with the air bags as original equipment. Here the role of Ford Motor Company was essential.

New Management, New Strategy

Lee Iacocca was fired as president of Ford Motor Company by Henry Ford II in July 1978. While Iacocca later moved over to lead Chrysler, the helm at Ford was taken by Philip Caldwell (chairman) and Donald Petersen (president and chief operating officer). Ford's progressive thinking on air bags arose from this new management team and from a new product philosophy that was enunciated by former Chairman Henry Ford II just prior to his retirement in 1982. Every aspect of the company's product was to be reexamined from the perspective of customer value.

In the 1970s Ford viewed safety as something everyone in the car business must do to satisfy regulation, that all manufacturers would suffer roughly equivalent cost penalties. Safety itself was not viewed as a source of customer value.[5] As he believed consumers were apathetic, Iacocca in particular wanted government to specify safety goals.[6]

In the 1980s Ford's new management began to look at safety as a product characteristic related to customer satisfaction. Marketing surveys showed that three product attributes were of growing importance to customers: reliability, security, and dependability. Although the demand for safety was not obvious, safety was interpreted to be an aspect of these desired attributes. There was also evidence that public attitudes toward safety were changing as reflected in child-restraint legislation, tough drunk-driving legislation, and new interest in mandatory belt use legislation.[7]

Unlike Iacocca and Henry Ford II, the new generation of management at Ford had no strong opinions against the air bag technology. Indeed, when Petersen was shown the detachable automatic belt designs that Ford had planned to use to comply with the Adams rule, he reportedly responded with "revulsion and loathing."[8] As a result, Ford's air bag advocates, veteran Vice President Herbert Misch and automotive safety director Roger Maugh, were given some freedom to make new development decisions on occupant restraints.

Misch and Maugh were strong believers in the air bag as a supplement to belt systems. During the 1981–1983 recession when the company's financial picture sank into the red, Misch found funds for a small but intensive air bag development program. And it was

Maugh who persuaded Ford's management to bid on the NHTSA-GSA demonstration contract.[9] According to career NHTSA officials, "he [Maugh] just wouldn't let it [the air bag] go away. There's no one at GM that has performed the same role."[10] Misch had remained "remarkably open to new ideas," despite two decades of fighting on the occupant-restraint issue.[11] Ford emerged as the only bidder on the air bag contract and was ultimately awarded $35 million to produce 5,000 compact sedans (Tempos) with supplemental driver-side air bags.[12]

Ford did not pursue the air bag simply because of the availability of the NHTSA-GSA contract. As early as October 1981, when (coincidentally) NHTSA announced rescission of the passive-restraint rule, Ford decided to revive its air bag research program. According to Ford Vice President Helen Petrauskas (Misch's successor), the air bag program was renewed after the occupant-restraint issue was reanalyzed as one of "customer satisfaction per se," not "regulatory compliance."[13] She emphasized, however, that the NHTSA-GSA program was a "marvelous opportunity" for Ford to accumulate real-world experience with the technology without incurring significant financial losses.[14] In this sense the NHTSA-GSA program made it easier for Ford to gain confidence in the device and expand its offerings to other fleet buyers and ultimately to the general public.

Petrauskas herself offered a fresh, innovative perspective on the safety issue. She had worked in environmental affairs at Ford for twelve years until she succeeded Herb Misch in mid-1983.[15] She had no background in the air bag wars of the 1970s and was determined to convert Ford from a "reactive to a proactive posture on the safety issue."[16] Her philosophy was that safety was not an issue of corporate altruism but a product attribute that customers would appreciate in the long run. She prodded Ford's top management toward air bags as a long-run investment in customer value, despite the short-run cost penalty.[17]

Several unexpected salespersons came forward to support Ford's fledgling program. Brian O'Neill, president of the Insurance Institute of Highway Safety, organized a large conference in Washington, D.C., where fleet buyers were urged to purchase air bag–equipped Tempos.[18] Peter Libassi, a senior vice president of Traveler's Insurance Company, made the decision to make the first large bid for a fleet of Ford's air bag cars.[19] The chief executive of USAA, General Robert McDermott, also supported O'Neill's efforts to promote the air bag to fleet buyers.[20] And even Ralph Nader is credited with persuading several fleet buyers to go for the Ford Tempos.[21]

Ford was not the first car manufacturer in the 1980s to offer air

bags to American consumers. Mercedes-Benz had been successfully offering supplemental driver-side air bags on its European
sales in the early 1980s. In February 1983 the company announced
that because of the advent of belt use laws in the United States, it
intended to offer driver-side air bags as an $880 option in North
America beginning with certain 1984 models.[22] BMW quickly responded with plans to offer air bags as an option in North America
on some 1985 models. BMW's offering came on the heels of a
highly publicized head-on crash in Europe between a BMW and a
Mercedes equipped with an air bag. The BMW driver was killed
while the Mercedes Driver walked away.[23] The offerings by
Mercedes-Benz and BMW were important indications of faith in air
bag technology, but they meant only that certain luxury buyers
would have the opportunity to own an air bag–equipped car.

As early as August 1984 there were indications that Ford was
preparing to extend its offering of driver-side air bags to the general
public. In a petition to NHTSA, Ford requested that NHTSA count
cars with driver-side air bags as one credit toward the automatic-
restraint quotas in Secretary Dole's plan. Such cars would have
received no credit under Dole's original plan since the passenger
side was not equipped with passive protection.[24]

The Ford proposal, which would apply only during the phase-in
period (model years 1987 through 1989), was endorsed by the Insurance Institute for Highway Safety, which was eager to encourage
air bag offerings. The IIHS also proposed that a two-car credit be
authorized for cars with full-front air bag systems. General Motors
Corporation also supported Ford's petition and recommended that
the credit for cars with driver-side air bags be extended beyond
model year 1990.[25] Despite the substantial support for Ford's petition, NHTSA was slow to respond, in part due to litigation of Dole's
original plan in the D.C. Circuit.

Competitive Pressure

The new air bag programs at Mercedes-Benz, BMW, and Ford
induced reassessments of corporate strategy throughout the industry. Was the interest in air bags a fad, or was it likely to grow? By
August 1985 there were news reports from anonymous sources that
air bag programs were also underway at Volvo, Honda, and General Motors.[26] At about this time, Mercedes-Benz announced that
driver-side air bags would be made standard equipment on some
1985 lines, and in October 1985 Mercedes-Benz announced that
driver-side air bags would be made standard equipment on all 1986

cars (90,000 projected sales).[27] Volvo then announced plans to offer driver-side air bags on certain 1987 models.

This flurry of interest in air bags occurred just prior to NHTSA's public response to Ford's petition. In the fall of 1985 NHTSA Administrator Diane Steed officially approved Ford's October 1984 petition to count cars with driver-side air bags as one credit toward the 1987–1989 compliance quotas. Full-front automatic protection through either air bags or friendly interiors would count as 1.5 credits. No special consideration was to be given to cars produced with air bags during model years 1990 or later.[28]

Steed's decision had significant business implications because it meant that a driver-side air bag system was equivalent to two automatic belts (driver and passenger) from the standpoint of credit toward compliance. The cost disadvantage of the air bag technology relative to automatic belts was thus substantially reduced, at least during the phase-in period. Meanwhile, the inability of engineers to design automatic belts that permit easy access to the front-center seating position in large cars was helping to fuel interest in the air bag.[29]

Ford responded to Steed's decision by announcing that supplemental driver-side air bags would be made available as an $815 option on both the 1986 Ford Tempo and the 1986 Mercury Topaz.[30] With this action, Ford became the first manufacturer to offer air bags on American-made cars since GM did so in the mid-1970s. But unlike GM's offerings on luxury models, the Tempo and Topaz were both modestly priced models ($7,508 and $8,235, respectively). Hence, Ford was the first manufacturer to offer air bags to middle-income buyers; the option was scheduled to begin on four-door sedans starting with March 1986 production. Because Ford had already produced 7,400 four-door Tempos (model year 1985) with air bags for GSA and various fleet buyers, the new optional air bag program was relatively easy to implement.

Much to the dismay of Ford officials, their carefully planned press announcement of the first air bag option was preempted several days in advance by a leak to the press. Apparently Ford had let air bag advocates, including Joan Claybrook, know a few days in advance that the official announcement was coming. Claybrook leaked (perhaps inadvertently) this information to the press and in effect preempted Ford's public announcement.[31] As a result, Ford Motor Company's distrust of Claybrook, which had begun in the 1970s, was aggravated.

Ford's new air bag offerings nevertheless received massive publicity in trade journals and much favorable mass media attention. Chrysler and General Motors soon surprised the industry by revealing plans to follow Ford's lead. The *Wall Street Journal* reported in December 1985 that Chrysler had contacted air bag suppliers and was looking to offer air bags on two 1987 compacts, the Dodge Daytona and the Chrysler Laser.[32] In February 1986 the *Journal* also reported that GM would offer driver-side air bags on selected 1988 models, though this commitment seemed less certain and extensive than the plans of Ford and Chrysler.[33] In a speech to the American Medical Association, Chairman Roger Smith explained the rationale for GM's new program: "We're offering this system in response to what we think our customers want rather than in response to a government mandate and that's the way it ought to be."[34]

European importers were also moving rapidly to expand their air bag offerings. BMW was planning sales of about 10,000 1987 cars with optional driver-side air bags. In early 1986 BMW also announced plans to upgrade air bags to standard equipment on the L7, a new luxury car with projected sales of 1,500. Volvo decided to offer driver-side air bags during model year 1987 in its top-of-the-line cars, and Porsche planned full-front air bag options in model year 1987.[35]

The luxury car market was especially suitable to the air bag because the front-center seating position of a six-seat passenger car could not be occupied easily when the driver and right-front positions were equipped with automatic belts. Marketing analysis also found that buyers reacted negatively to some automatic belt designs, citing discomfort, inconvenience, and unsightliness. As soon as it became apparent that the driver-side air bag system was not much more costly than full-front motorized belts, the phones of air bag suppliers were ringing. As one supplier put it, after years of dormancy the air bag industry "is going quite nuts."[36]

The only mystery in this competitive scramble was the planning of Japanese manufacturers, which were characteristically silent about their compliance plans. News reports did indicate that Honda was thinking about air bags for the new Legend.[37] But it seemed apparent that the Americans and Europeans were moving more rapidly to the air bag than were the Japanese.

An Unexpected Alliance

In early 1986 each car manufacturer was faced with automatic restraint quotas of 10 percent in 1987, 25 percent in 1988, 40 percent in 1989, and 100 percent in model years 1990 and later. Ford's legal experts persuaded top management that the passive-restraint standard was not likely to be rescinded, placing great weight on the reasoning of the Supreme Court's 1983 decision.[38] Even if DOT tried to activate the "trapdoor" in April 1989 (which was considered unlikely), another round of litigation was certain. Roger Maugh of Ford explains: "We believed that automatic restraints would be required one way or the other in the long run."[39]

As Ford's automotive safety director, Roger Maugh wanted to move toward air bags as Ford's long-run compliance technology. Maugh saw safety belts as the lifesaver and air bags as a source of supplemental crash protection to mitigate potential chest injuries and help prevent facial disfigurement. He was supported by Vice President Helen Petrauskas. While Ford's driver-side air bag technology was established, Maugh was still working with TRW, Inc., an air bag supplier, on a safe passenger-side air bag. Maugh was concerned about injury to out-of-position passengers (especially standing children). He knew Ford's passenger-side bag would not be ready until at least the 1989 model year. Indeed, he was not yet certain that a satisfactory passenger-side system could be developed.

The phase-in period (1987–1989) in NHTSA's rule did not pose an immediate problem for Ford because NHTSA was treating a driver-side air bag as equivalent to full-front automatic belts from a standpoint of regulatory compliance. Yet the investments Ford made during the phase-in period would determine which restraint system would be ready for post-1989 model cars. The easy course, which was being followed by both GM and Chrysler, was to go primarily with automatic belts and provide only limited air bag offerings. Likewise, Ford could delay indefinitely the introduction of air bags as standard equipment until the passenger-side system was ready. Automatic belts could be used in the interim, but this strategy would require that Ford invest scarce resources in the development of both types of automatic systems. Alternatively, Ford could go with driver-side air bags and gamble that the passenger-side systems would be ready by model year 1990 (the year the extra credit for driver side air bags was scheduled to expire). The Dole plan made model year 1990 a double challenge for manufacturers because of the large increase in fleet coverage (40 to 100 percent) and the extension of passive protection to passengers as well as drivers.

Maugh and Petrauskas did not like any of these options. In conversations with NHTSA's top personnel, Ford officials tried to gauge the agency's receptivity to a possible extension of the credit into the early 1990s. They quickly gathered that this proposal was politically infeasible without support from air bag advocates.[40]

Ford did not give up. Petrauskas and Maugh decided to try to enlist support from the Insurance Institute for Highway Safety. "Of all the safety groups," Petrauskas explains, "we felt IIHS would give us the fairest hearing."[41] Maugh arranged a one-on-one meeting in March 1986 with Brian O'Neill, president of the Insurance Institute for Highway Safety. This was a remarkable initiative considering that Ford and IIHS had been bitter antagonists on numerous safety issues since the early 1970s. The youthful Petrauskas recalls: "We wanted to control our own destiny. I saw no reason why we couldn't or shouldn't talk to IIHS."[42] Moreover, Maugh knew that O'Neill was a dedicated air bag advocate. Yet unlike his predecessor at IIHS (William Haddon), O'Neill was perceived as a pragmatist.

At their one-on-one meeting, Maugh told O'Neill that the 1990 expiration of credit for driver-side air bags would "kill the air bag at Ford."[43] Without an extension of the credit, Ford would be compelled to abandon air bags in favor of automatic belts in all except Ford's luxury lines. Maugh emphasized that Ford "could not risk the stockholders' interests on theoretical passenger air bags."[44] Maugh also presented calculations showing that driver-side air bags—combined with improved belt-wearing rates—would save more lives than full-front automatic belts, a calculation that O'Neill confirmed independently. If Ford's petition was accepted, Maugh told O'Neill, driver-side air bags would be made standard equipment on a majority of Ford's lines by the late 1980s. Maugh concluded by requesting that IIHS endorse Ford's planned petition to NHTSA to extend the driver-side credit beyond 1990.[45]

O'Neill was a close observer of the air bag business and knew that no company (domestic or foreign) had plans to offer air bags outside their luxury lines. He saw Maugh's plan as a potential breakthrough for the air bag technology, something he and Haddon had worked for at IIHS since the late 1960s. Most importantly, O'Neill trusted Maugh.

O'Neill assured Maugh that IIHS would support Ford's planned petition. O'Neill also invited Maugh's boss, Vice President Helen Petrauskas, to present Ford's case to the IIHS Board in May 1986 in order to increase support among individual insurers.

Petrauskas recalls being "nervous before the meeting in San Antonio, wondering whether she had been served up for lunch."[46] She explained the three aspects of Ford's proposal: the business rationale, the safety consequences, and the broader public interest. She recalls answering some "very tough questions" about what "assurances" there were. She emphasized that Ford could not pursue both air bags and automatic belts "due to a shortage of high-quality restraint engineers."[47] The result of the meeting (and some gentle lobbying by O'Neill) was that the insurance industry generally became enthusiastic supporters of Ford's upcoming petition.

O'Neill, at Maugh's request, elected not to consult with Joan Claybrook and Clarence Ditlow of the Center for Auto Safety until a few days before Ford's planned petition was about to be filed. Ford feared a preemptive move by Claybrook. Maugh, in particular, did not trust Claybrook. As we shall see, this secrecy later caused bitterness between Claybrook and O'Neill.[48] As mentioned earlier, apparently Claybrook had leaked to the press (perhaps inadvertently) advance knowledge of Ford's earlier announcement that air bags would be offered for sale on the Ford Tempo and Topaz.[49]

The Cold Reality of the Marketplace

Based on the publications of professional economists and marketing analysts, one might think it would be fairly easy to sell air bags to consumers. For example, a 1986 study by economists at the Brookings Institution found that mandatory air bags would have estimated benefits ($19.2 billion per year) far in excess of their estimated costs ($4.2 billion per year).[50] The estimated benefits included fewer lost work days, less care for injured motorists, and the presumed willingness of consumers to pay money to avoid pain and suffering.

Marketing studies and opinion polls also indicated substantial consumer support for air bags. A 1986 survey of 15,475 new-car buyers by *Newsweek* magazine found 87 percent of respondents in favor of air bags as standard equipment. Even when told about the potentially high price of air bag systems, substantial percentages of respondents indicated a preference for air bags.[51] When citizens were told that mandatory air bags would add $350 to the price of a car, a majority (52 percent in one survey) favored the new safety system. When asked whether they would buy an air bag system at prices of $500 and $1,000, the percentage of consumers in one survey who said "definitely buy" or "probably buy" was 33 and 17, respectively. The results of this survey by economists at Northeastern University seemed to suggest that even high-priced optional air bags would be salable in some segments of the new-car market.[52]

The showroom reality was starkly different. Ford was not successful in selling cars with optional driver-side air bags. As of February 20, 1987, Ford reported that 455 Tempos and 294 Topaz cars with air bags were sold to retail buyers out of a total sales rate of 100,000 for the two models. An estimated 2,833 air bag cars were sitting unsold in dealer inventory, an amount equal to an 18-month supply at the historical sales rate.[53] By May 1, 1987, after intensified marketing efforts, the retail sales count had improved only modestly: 1,014 Tempos and 439 Topazes.[54] The slow sales in model year 1986 were somewhat expected due to the option being introduced late in the model year. But the order books during model year 1987 were also remarkably empty. Before concluding that car buyers just don't want air bags, it is important to consider how the new-car market actually works. A Tempo GL two-door in model year 1986 listed for $7,358. The choice of an optional air bag ($815) did not dictate a final sales price of $8,173, however, for the air bag was made available by Ford only on the Tempo four-door, base price $7,508. For cars ordered from the factory, a destination charge of $398 might have been added. To buy the four-door Tempo with the air bag, the buyer also had to purchase several other options: power steering ($223), automatic transmission ($448), and air conditioning ($743). As the *Boston Globe* reported in March 1986, an air bag–equipped Tempo really cost consumers about $10,135.[55]

Nor were dealers always enthusiastic about marketing the air bag option. Ralph Nader's Center for the Study of Responsive Law surveyed forty-five dealers in fifteen states in 1987 to determine whether air bags were available on the Tempo. Only thirteen said it was, and according to study leader James Musselman, most of the other dealers did not want to find out. None of the surveyed dealers mentioned the air bag when first discussing options.[56] One interpretation of this finding is that dealers knew consumer desires and made an informed professional judgment that the air bag option is a loser. David Martin of GM, for example, saw dealer reactions as a clear indication that consumer demand for air bags is minimal.[57]

Although dealers insist that consumers are the biggest critics of air bags, the *Wall Street Journal* reported some other explanations for dealer resistance. Because few 1986 Tempos in stock were equipped with air bags, they had to be specially ordered. Yet dealers like to sell cars already on the lot. Frank McCarthy, executive vice president for the National Automobile Dealers Association, emphasized that some dealers also feared that they would be sued if an air bag malfunctioned in an accident. According to media consultant Ben Kelley, the challenge for Ford was to "reteach its dealers that this thing the auto industry said for 15 years was so harmful is now so great."[58]

In model year 1987 Ford's management made a concerted effort to sell the air bag option. Chairman Donald Petersen sent all 6,000 Ford and Lincoln-Mercury dealers a letter stressing that dealer support was "critical" to the success of the air bag option. About 500,000 pieces of pro–air bag literature were sent to dealers, and Ford's safety director made personal visits to numerous dealers to push the air bag.[59]

In mid-February 1987 a group of Ford and Lincoln-Mercury dealers was convened by the top management of Ford to discuss ways to improve public acceptance of the air bag. The major impediments to sales that were cited by dealers included:

- Price ($815).
- Lack of marketing support except for point-of-purchase displays and dealer liability for servicing/replacement.
- Lack of consumer interest in safety.
- The attractive set of options that cannot be included with the air bag in model year 1987 (manual transmission, a factory-installed tilt wheel, and cruise control).

Ford's management responded to the low sales rate by temporarily cutting the price of the option from $815 to $295. This was accomplished by offering buyers a $520 cash rebate for taking cars with air bags.[60]

General Motors had a similarly poor sales experience with optional air bags in the 1974 model year. But unlike GM, which abandoned the technology (while downsizing their lines) after an initially poor marketing record, Ford persisted in its program. According to one Ford spokesman: "In our view, air bags are simply too important to be judged on the basis of slow retail sales. We have made a commitment to the government."[61] Indeed Ford was moving toward a major commitment to driver-side air bags as standard equipment on 1990 and later models.

Short-Run Compliance Strategy

Although air bags were central to Ford's long-range plans, the company was faced with imminent automatic-restraint quotas of 10 percent, 25 percent, and 40 percent of its fleet for model years 1987, 1988, and 1989, respectively. Failure to meet these quotas would mean adverse publicity and fines of $1,000 (negotiable) per violation. A driver-side air bag system was not yet ready for every line. Practically speaking, Ford had to invest in some type of automatic belt system to help satisfy the 1987–1989 quotas.

For years most automakers had expressed a preference for nonmotorized over motorized automatic belts. The nonmotorized systems were easily detached and cheap, while the motorized systems were essentially nondetachable and expensive. For example, the motorized design on the Toyota Cressida had achieved over 95 percent usage rates but cost $350 per car (in low-volume production).[62] As the 1987 model year approached, Ford's marketing department made an extensive study of the design choices.

In September 1986 Ford surprised the industry by announcing plans to install motorized shoulder belts on all 1987 models of its Escort and Mercury Lynx cars, starting with December production.[63] Ford's plan was to license the Cressida system technology from Toyota.

Ford's top management spurned the cheaper nonmotorized systems in part on the basis of marketing tests in which consumers reacted more favorably to the motorized design. In-house skeptics at Ford questioned this decision on the grounds that the tests were conducted primarily on luxury cars. They also argued that it would be easier to convert back to manual systems from nonmotorized automatic designs if NHTSA ultimately rescinded the passive-restraint standard. Despite these arguments, Ford's management surprised the industry by choosing the motorized design. According to Ford Chairman Donald Petersen, the motorized system is "more expensive but, to us, the benefits warrant the incremental cost."[64]

Ford was ultimately not alone in choosing a motorized design for model year 1987. They were joined in this decision by Nissan, Saab, Mazda, Mitsubishi, Subaru, Ferrari, Alfa Romeo, and Isuzu. And Toyota installed its motorized system in the redesigned Camry as well as the Cressida.

In contrast, GM chose to satisfy the 10 percent quota for model year 1987 by making nonmotorized three-point passive belts standard on its N- and H-body cars. Volkswagen decided to offer its original nonmotorized automatic shoulder belt on most Golf and Jetta models but not on the Audi.[65] As a result, new-car buyers were offered a considerable diversity of automatic belt designs.

Insurers and Ford Go Public

Knowing that Secretary Dole and NHTSA officials were eager to promote air bags, Ford officials placed their long-run dilemma (motorized belts versus air bags) before the government. In particular, they petitioned NHTSA officials in June 1986 to revise their rule so that supplemental driver-side air bags and passenger-side manual belts would continue to satisfy the standard in 1990 and later models. In exchange for this amendment, Ford made the startling promise that it would "in all likelihood" make driver-side air bags standard equipment on a majority of its post-1990 North American cars. If the petition were denied, Ford intimated that it would instead devote its energies to installing automatic belts in both front-seat positions.[66]

The Ford petition led to an extremely unusual and tense split within the auto safety community. Joan Claybrook of Public Citizen and Clarence Ditlow of the Center for Auto Safety denounced the Ford proposal. They were supported, at least initially, by Senator John Danforth (R–Mo.) and Representative Timothy Wirth (D–Colo.). But in an uncommon alliance, insurers led by IIHS President Brian O'Neill joined most vehicle manufacturers in support of Ford's petition.

Clarence Ditlow, head of Ralph Nader's Center for Automobile Safety, reacted as follows: "They want to protect the driver but not the passenger—which tells you what I think about their petition. It would be opening the floodgates for amendments and would further delay getting air bags into cars."[67] Congressman Wirth concurred, stating that this is "just the latest effort by Ford to keep full-front air bags from the general public."[68] Senator Danforth's initial reaction was also critical:[69]

> The "better idea" people at Ford must not have been involved in developing the Company's passive restraint proposal. It indefinitely delays full front seat automatic crash protection and provides no guarantees that lifesaving air bags will be more widely available.

Danforth urged NHTSA to reject or radically alter the proposal.

The reaction of the insurance industry to Ford's proposal was organized by Brian O'Neill, president of the Insurance Institute for Highway Safety. In a public statement, O'Neill stressed that "Ford's promise to produce large numbers of cars with air bags" is a "breakthrough" toward making the safety device standard equipment "at a reasonable price."[70] In an analysis submitted to NHTSA, O'Neill concluded that 70 percent of all front-seat occupants would have to wear automatic belts for those cars to equal the lifesaving protection offered by driver-side air bags.[71] IIHS did recommend that the deadline for full-front passive protection be relaxed only until model year 1994 rather than be postponed indefinitely. Following the lead of IIHS, Ford's petition received favorable comments from Traveler's, State Farm, Allstate, Aetna, Nationwide, and USAA.

To defuse opposition to Ford's petition, O'Neill suggested a meeting of the National Coalition to Reduce Car Crash Injuries (a consortium of medical groups, crash victims, air bag suppliers, Naderites, and insurers) to hear a presentation by officials of Ford

Motor Company. After a tense session in which the Ford team attempted to answer a series of tough and skeptical questions, the Coalition members—except Joan Claybrook and Ditlow—elected to support Ford's position. Claybrook and Ditlow insisted that Ford should commit to a phase-in of passenger-side bags over the late 1980s and early 1990s.[72]

At a later lunch with O'Neill, Claybrook and Ditlow questioned O'Neill's failure to consult them before siding with Ford. They argued that if O'Neill allowed Ford to split insurers from consumer advocates, the long-run effect would be to cripple the auto safety movement. O'Neill responded that any further conditions on Ford's petition would only make it less likely that Ford would go with air bags. He felt that in the long run driver-side air bags would bring passenger-side air bags.[73]

The issue came sharply into focus in December 1986 at public hearings before Senator Danforth's Commerce Committee. Ford Vice President Helen Petrauskas explained to Danforth that "the Ford proposal does not weaken the passive-restraint standard. It makes no change unless a manufacturer installs air bags."[74] If the petition were granted, she stressed, Ford would offer driver-side air bags in 500,000 to 1 million cars in model year 1990. When asked by Danforth whether her testimony constituted a "company commitment," Petrauskas responded in the affirmative.[75]

At the same hearing Chrysler and General Motors also supported Ford's petition. When asked by Danforth how Chrysler would respond, Christopher Kennedy stated that driver-side air bags would be installed in about 500,000 Chrysler cars in model year 1990 if the proposal is approved by NHTSA. General Motors also testified before Danforth but refused to make any specific commitments beyond their earlier plan to offer air bags on about 10,000 of its $13,500 Oldsmobile 88s.[76]

Senator Danforth was encouraged by the testimony at the hearing. Only Clarence Ditlow of the Center for Automotive Safety was critical of Ford's proposal, yet his remarks were neutralized by Brian O'Neill's favorable testimony. After hearing Ford's commitment to the air bag technology, Danforth revised his earlier view and concluded that "granting the petition is desirable."[77]

Roger Smith's Turnaround

While Ford and Chrysler were launching large-scale air bag programs, GM Chairman Roger Smith was much more tentative. According to one view, the air bag issue had become "freighted with

political, symbolic, facesaving value for Smith."[78] When GM spokes-
men were queried on the matter, their responses were highly
guarded. Twice in 1985 the CBS program "60 Minutes" ran highly
favorable stories about the air bag technology, but GM refused to be
interviewed. The son of the late Ed Cole, David Cole, a knowledgeable observer of the auto industry, explained that the air bag had
"become a very touchy, emotional subject" at General Motors.[79]

Despite Smith's hostile feelings about the air bag issue, he saw
that the company's position had to change. Several weeks prior to
NHTSA's final decision on Ford's proposal, General Motors
switched course and announced a major commitment to air bags
as standard equipment on 1990 and later cars. Assuming Ford's
petition were granted, GM Chairman Roger Smith announced
plans to install air bags on 500,000 1990 cars and three million
1992 cars. Smith attributed the new offerings to a "new interest in
safety" on the part of the customers.[80]

The decision by GM came as a shock to many auto experts who
thought Smith would never approve a large air bag program. GM
insiders explain that the company moved slowly toward air bags in
the 1980s because management felt that Ford's interest might
prove to be a "flash in the pan."[81] Many GM officials remembered
the company's optional air bag program in 1974–1976, which was
widely perceived as a financial loser.

The issue at GM pitted those who viewed air bags as a bad
business investment against those who saw air bags both as a necessary step to meet competition and a favorable step toward promoting the company's image for social responsibility. Historically,
Smith sided with the former viewpoint, perhaps reflecting his background as a "bean counter" and his perception of air bags as a
regulatory problem. As competition intensified, Smith's opposition
softened. Since competitive pressure for air bags was greatest in
the luxury lines, that is where Smith softened first. Ultimately,
he—like Iacocca at Chrysler—approved a large-scale air bag program to match Ford's initiative. Roger Smith explains:[82]

> We are motivated by what we perceive to be a more positive attitude
> today concerning air bag technology, where the inflatable restraint
> is used in conjunction with belts. This did not exist when GM pio-
> neered this concept during the 1974–76 model years. We believe the
> supplemental air bag technology has come of age.

The Dissenters

In March 1987 Secretary of Transportation Elizabeth Dole adopted
Ford's petition with some minor revisions. She delayed full-front

passive protection until model year 1994 for those cars in which air bags (i.e., nonbelt passive protection) were installed on the driver side. The safety of passenger-side manual belts was strengthened by requiring that such belts be dynamically tested.[83] Dole explained her decision: "The action we are taking will result in the installation of more air bags sooner than would have occurred without this rule. It will also encourage the orderly development and production of passenger-side air bag systems."[84]

Joan Claybrook denounced Dole's decision as "unconscionable" and later urged NHTSA to reconsider the decision. She pleaded with NHTSA not to accept "more promises from an industry that for almost twenty years has had the option of using air bags, but has defiantly refused to do so."[85] Ditlow joined Claybrook, and their petition for reconsideration made it clear that another lawsuit was in the making.

One of the fascinating aspects of this episode was the virtual silence of Ralph Nader, who made few public comments about Ford's petition. One interpretation is that Nader was supporting the positions of Claybrook and Ditlow by letting them take the lead.[86] Roger Maugh of Ford speculates otherwise: "Ralph Nader never vigorously opposed our petition. He was pragmatic about it. Claybrook though was so emotionally wrapped up she couldn't change."[87]

Claybrook offers a different interpretation of her reaction. She felt that Ford was bluffing and that a "brinksmanship" approach would have brought full-front air bags by 1990. In her words, "O'Neill fell for the trap." She urged NHTSA to order Ford to release their contracts with air bag suppliers in order to determine whether passenger-side air bags were already on order.[88] NHTSA Administrator Diane Steed denied this request.[89]

Claybrook responded by recruiting a coalition of consumer groups to sue NHTSA for "arbitrary and capricious" behavior in the denial of her petition for reconsideration. Claybrook reasoned that it was important—in the event that Detroit reneged on its end of the deal with NHTSA—for the consumer movement to be on record against the IIHS-Ford-NHTSA alliance.[90] Though the prospects of winning the lawsuit were not great, Claybrook saw an important principle at stake.

Iacocca Steals the Headlines

In its low-key fashion, Ford Motor Company was trying to use its progressive air bag program as a technique to rise above its competitors in the eyes of consumers and opinion leaders. Early in 1987

Ford continued this strategy by announcing its plans to be the first domestic manufacturer in the 1980s to offer passenger-side air bags in addition to the driver-side system. Ford's plan was to offer them as an option on some 1989 models.[91]

Meanwhile, insurers were taking steps to make air bags more attractive to both dealers and car buyers. In the spring of 1988, anticipating a move by USAA, State Farm announced a new program of medical premium discounts: 10 percent reduction for cars with automatic belts, 20 percent for driver-side air bags, and 40 percent for full-front seat air bags and automatic belts.[92] The State Farm program mirrored a program that Nationwide had been offering since 1986. Chairman Robert F. McDermott of USAA saw State Farm's announcement the day before his scheduled March 30 Washington news conference as a preemptive move, but he was not deterred. He held the press conference as scheduled, announcing a doubling in the premium discounts for air bags on certain policies, from 30 to 60 percent (or about $30 to $40 per year in savings for the average policyholder). In addition, McDermott announced that a $300 bonus would be awarded to policyholders who purchased air bag–equipped cars. Incentives were also offered for GM and Ford dealers who sold large numbers of cars with air bags. McDermott explains: "What we're offering is to pay for that [air bag] for you."[93]

Few business executives in America are more adept at using the mass media than Lee Iacocca. He saw that Ford was moving to capture the prosafety image in the auto market and that insurers were determined to make air bags a reality. While Chrysler had already announced its decision to go with air bags, Iacocca had not yet used the air bag as a mass media ploy. After all, knowledgeable industry observers knew that "Iacocca was probably the auto industry executive most identified with opposition to air bags."[94]

Early in 1988 there were reports in the industry that Chrysler might be the first domestic automaker to offer driver-side air bags as standard equipment.[95] Iacocca decided to use this fact as a counterpunch against Ford in a battle for the prosafety image. At a major press conference in May 1988, Iacocca announced that Chrysler would be first. His announcement was covered by all the major TV news networks. Accompanying his press announcement were full-page ads in the *New York Times* and other papers saying:[96]

> *Who says you can't teach an old dog new tricks?*
>
> *I had my doubts about air bags, but today's new technology made me a believer.*

The *New York Times* billed the Iacocca decision as "signaling the end of a nearly 20-year battle over use of the devices."[97] Ralph

Nader, who had been "pretty much on the sidelines" throughout the 1980s, used the Iacocca announcement to his own advantage.[98] He told the *New York Times:* "We have declared victory; this is a remarkable turnaround."[99]

The Last Chapter

Despite Iacocca's announcement and Nader's declaration of victory, a few loose ends in the story still had not been told. The future of passenger-side air bags was still a bit cloudy, Joan Claybrook's lawsuit against NHTSA was still before the D.C. Circuit Court of Appeals, and the precise extent of GM's air bag effort was still in question.

Ford Motor Company was not about to let Iacocca steal the limelight on air bags without a fight. In June 1988 Vice President Helen Petrauskas came to Washington to announce that Ford Motor Company would be the first American carmaker to make both driver- and passenger-side air bags standard equipment. After a successful media demonstration of the twin air bags, Petrauskas explained to the press that Ford was determined to see to it that the passenger-side air bag was fully developed.[100]

Meanwhile, at General Motors, the air bag program was moving forward but still with a degree of hesitancy not present at Chrysler and Ford. GM President Robert Stempel announced in July 1988 that air bags would be offered as standard equipment on 500,000 of its 1990 model U.S.-made cars. While that number may seem huge, it represents only about one-seventh of GM's annual production. Stempel explained: "We're not going to put air bags on if people don't want them. If I can get the same kind of protection without air bags, that's the way we ought to go."[101] GM analysts were still concerned that consumers might not be willing to pay the price of air bags. Stempel emphasized that GM's $850 air bag option on the near-luxury Oldsmobile Delta 88 "is not selling like the proverbial hot cakes." Stempel was especially concerned that air bags might prove to be too expensive for economy cars.[102]

The "final" line of the story occurred on July 15, 1988, when a unanimous three-judge panel of the D.C. Circuit ruled on the legality of Secretary Dole's phase-in period for passenger-side air bags. Although a loss by NHTSA in this case seemed unlikely, it would have sent a wave of regulatory uncertainty through Detroit and complicated the completion of the bustling air bag development programs.

The court affirmed NHTSA's decision to grant a four-year extension as "a reasonable approach toward increasing the nonmarket incentives for automakers to install air bags rather than automatic belts in future car models."[103] The court was not, however, completely hostile to Claybrook's perspective. The final line of the opinion reads: "While we do not read NHTSA's decision as precluding absolutely any further extension, we do take its reasoning to indicate a strong presumption against any such suspension."[104]

After this court case was decided, only one mystery remained. It is a mystery that permeated the entire twenty-year struggle. What were the Japanese manufacturers doing on air bags? They were "the only major manufacturers without public plans to make the life-saving technology of air bags widely available."[105]

Conclusion

A key lesson from this 1985–1988 period is that market forces, if harnessed, are a faster and more effective road to safety than technology-forcing regulation. That does not mean that regulation is worthless or unnecessary. Market interest in safety is highly sensitive to managerial personalities as a result of latent and fickle consumer demand for safety. The challenge for regulators is to foster an environment in which manufacturers perceive that consumers are interested in safety and that any firm that does not provide it will be placed at a competitive disadvantage. As we have seen, Elizabeth Dole was successful in creating just this type of environment.

Dole's technology-inducing strategy would probably not have worked without the progressive corporate strategy followed by Ford Motor Company. Vice President Helen Petrauskas, in particular, saw the air bag issue as an opportunity for Ford to regain control of its destiny in the safety area. She accomplished what few auto executives had contemplated for the previous twenty years: a pro-regulation deal with the insurance industry that promised to bring air bags to the marketplace. Although it is still too early to say for certain whether Ford's new strategy was competitively successful, there is no doubt that Ford—not GM, Chrysler, or the Japanese—was controlling its destiny in Washington through an innovative corporate strategy.

Endnotes

1. "NHTSA Eyes Two Fleet Tests of Air Bags," *Automotive News* (December 18, 1982); "Arizona Company Wins NHTSA Contract for Retrofitting Air Bags," *Status Report: Highway Loss Reduction* (April 22, 1983), p. 1.

2. Testimony of NHTSA Administrator Diane Steed, *DOT and Related Agencies Appropriations for 1986,* House Committee on Appropriations, 99th Congress, First Session, February 1985, p. 40.
3. "Breed Air Bag System Passes NHTSA Tests," *Status Report: Highway Loss Reduction* (February 16, 1985), p. 1.
4. "GSA Signs Agreement for Air Bag Fleet," *Status Report: Highway Loss Reduction* (April 22, 1983), p. 8; "Ford Expected to Supply Air Bag Cars for U.S. Test," *Automotive News* (March 16, 1983), pp. 8, 45.
5. Interview with Helen Petrauskas, September 3, 1987.
6. Lee Iacocca, *Iacocca: An Autobiography* (Toronto: Bantam Books), 1984, p. 297.
7. Interview of Helen Petrauskas, September 3, 1987.
8. Interview of Don McGuigan, September 3, 1987.
9. "Ford Offers 10,000 Autos with Air Bags in Bid on U.S. Contract for Fleet Vehicles," *Wall Street Journal,* August 25, 1983, p. 2; *Status of Two DOT Air Bag Projects,* U.S. General Accounting Office, Washington, D.C., September 10, 1984, RCED-84-177.
10. "Ford and GM: Contrasting Styles," *Status Report: Highway Loss Reduction* (November 8, 1986), p. 6.
11. Interview of Helen Petrauskas, September 3, 1987.
12. "Federal Agency to Buy 5,000 Tempos Equipped with Air Bags," *Status Report: Highway Loss Reduction* (March 3, 1984), pp. 1–2.
13. Interview of Helen Petrauskas, September 3, 1987.
14. Ibid.
15. "Ford Elects Petrauskas as First Woman VP," *Automotive News* (March 16, 1983), p. 8.
16. Interview of Helen Petrauskas, September 3, 1987.
17. Ibid.
18. Interview of Brian O'Neill, July 24, 1987.
19. Ibid.
20. "Hearing Told USAA Will Buy 150 Tempos Equipped with Air Bags," *Status Report: Highway Loss Reduction* (March 24, 1984), pp. 1–2; "Transforming a Military Insurer into a Civilian Powerhouse," *New York Times* (June 28, 1987), p. F6.
21. Interview of Roger Maugh, July 8, 1987.
22. "Ford Offers 10,000 Autos with Air Bags in Bid on U.S. Contract for Fleet Vehicles," p. 2.
23. Ibid.; also, Interview of Carl Nash, July 23, 1987.
24. "Institute Urges NHTSA to Grant Ford Request on Restraint Rule," *Status Report: Highway Loss Reduction* (July 6, 1985), pp. 2–3.
25. Ibid.
26. "Successful Air Bag Programs Spark Auto Makers' Interests," *Status Report: Highway Loss Reduction* (August 10, 1985), pp. 1–2.
27. "Air Bags in All Mercedes," *Status Report: Highway Loss Reduction* (October 5, 1985), p. 2.
28. "NHTSA Agrees to Credits for Cars Equipped with Air Bags," *Status Report: Highway Loss Reduction* (October 5, 1985), pp. 6–7.
29. Interview of Roger Maugh, July 8, 1987.

30. "Ford to Offer Air Bags as Option on '86 Compacts," *New York Times* (November 3, 1985), p. 37.
31. Interview of Brian O'Neill, March 3, 1988.
32. "Chrysler Appears to Weigh Use of Airbags in Two Models," *Wall Street Journal* (December 12, 1985), p. 6.
33. "GM Will Offer Air Bags as Option on Some Models," *Wall Street Journal* (February 21, 1986), p. 6.
34. Ibid.
35. "Auto Industry Tools Up for Air Bags," *Status Report: Highway Loss Reduction* (December 7, 1985), pp. 1, 6.
36. Ibid., p. 1.
37. Ibid., p. 6.
38. Interview of Don McGuigan, September 3, 1987.
39. Interview of Roger Maugh, July 8, 1987.
40. Interview of Roger Maugh, July 8, 1987; interview of Helen Petrauskas, September 3, 1987.
41. Interview of Helen Petrauskas, September 3, 1987.
42. Ibid.
43. Interview of Roger Maugh, July 8, 1987; interview of Brian O'Neill, July 24, 1987.
44. Ibid.
45. Ibid.
46. Interview of Helen Petrauskas, September 3, 1987.
47. Ibid.
48. Interview of Joan Claybrook, July 24, 1987.
49. Interview of Brian O'Neill, July 24, 1987.
50. Robert W. Crandall, Howard K. Gruenspecht, Theodore E. Keeler, and Lester B. Lave, *Regulating the Automobile* (Washington, D.C.: Brookings Institution, 1986), pp. 82–84; "Brookings Report Questions Benefits of Auto Standards," *Wall Street Journal* (March 18, 1986), p.8.
51. "Safety Not First with Reagan, Nader Charges," *Automotive News* (December 22, 1986), p. 3.
52. "Auto Safety," *Wall Street Journal* (February 6, 1986), p. 27.
53. "Air-Bag Fords Lure Only 749 Retail Customers," *Automotive News* (May 18, 1987), p. 6.
54. "Ford Offers $520 Rebate on Air Bags," *Automotive News* (May 18, 1987), p. 6.
55. "There's More to Price than Sticker Shows," *Boston Globe* (March 15, 1986), p. 55.
56. "Auto Shoppers Encounter Stiff Resistance When Seeking Air Bags at Ford Dealers," *Wall Street Journal* (July 31, 1986), p. 27.
57. Interview of David Martin, September 3, 1987.
58. "Auto Shoppers Encounter Stiff Resistance When Seeking Air Bags at Ford Dealers," p. 27.
59. Ibid.
60. "Ford Motor Deflates Price for Its Optional Air Bags," *Wall Street Journal* (May 13, 1987), p. 22; "Ford Offers $520 Rebate on Air Bags," *Automotive News* (May 18, 1987), p. 6.

61. "Air-Bag Fords Lure Only 749 Retail Customers," *Automotive News* (March 2, 1987), p. 1.
62. *Automatic Safety Belt Usage in 1981 Toyotas*, U.S. Dept. of Transportation, NHTSA, February 1982, DOT-HS-806-146.
63. "Ford Plans to Offer Motorized Seat Belts on Some 1987 Cars," *Wall Street Journal* (September 4, 1986), p. 16.
64. "Petersen Ranks Passive Belts High on Ford's Safety List," *Automotive News* (September 8, 1986), p. 6.
65. "Some Wait to Decide on Passive Restraints," *Automotive News* (September 15, 1986), pp. 1, 9.
66. "Ford Asks U.S. to Defer Enforcing Rule on Air Bags," *Wall Street Journal* (June 13, 1986), p. 9; "Ford Seeking a Change on Passive Restraints," *Automotive News* (June 16, 1986), p. 6.
67. "Ford Asks U.S. to Defer Enforcing Rule on Air Bags," p. 9.
68. "Ford Petition Would Allow Air Bags on a Large Number of Cars," *Status Report: Highway Loss Reduction* (June 28, 1986), pp. 1, 2.
69. Letter from Senator John Danforth to Secretary Elizabeth Dole, August 15, 1986.
70. "Ford Petition Would Allow Air Bags on a Large Number of Cars," p. 1.
71. "Easing Phase-In Could Save More Lives," *Status Report: Highway Loss Reduction* (June 28, 1986), pp. 2–3.
72. Interview of Brian O'Neill, July 24, 1987; interview of Joan Claybrook, July 24, 1987.
73. Ibid.
74. "Ford Proposal on Air Bags Reviewed at Senate Hearing," *Automotive News* (January 5, 1987), p. 28.
75. "Ford and Chrysler to Install Air Bags if U.S. Alters Rule," *Wall Street Journal* (December 12, 1986), p. 25.
76. Ibid.
77. "Strong Air Bag Commitment," *Status Report: Highway Loss Reduction* (January 24, 1987), pp. 6–7.
78. Ralph Nader and William Taylor, *The Big Boys: Power and Position in American Business* (New York: Pantheon Books, 1986), p. 139.
79. Ibid., p. 141.
80. "GM: 3 Million Air Bags in '92," *Automotive News* (March 9, 1987), pp. 1, 55.
81. Interview of Robert Rogers, August 7, 1987.
82. Ed Janicki, "Air Bags: Mandated Market," *Automotive News* (February 29, 1988), p. E4.
83. "Makers Get 4-Year Delay on Passive Restraints," *Automotive News* (March 30, 1987), p. 58.
84. "NHTSA Decision Will Mean 'More Air Bags Sooner,'" *Status Report: Highway Loss Reduction* (April 11, 1987), p. 1.
85. "Nader Challenge May Renew Bag-or-Belt Uncertainty," *Automotive News* (May 4, 1987), p. 4.
86. Interview of Brian O'Neill, July 24, 1987.
87. Interview of Roger Maugh, July 8, 1987.
88. Interview of Joan Claybrook, July 24, 1987.
89. *Federal Register*, vol. 52, November 5, 1987, pp. 42440–42445.

90. Interview of Joan Claybrook, July 24, 1987.
91. "Ford Plans Air-Bag Option for Front-Seat Passengers," *Wall Street Journal* (May 29, 1987), p. 4.
92. "Some Insurers to Offer Discounts for Air Bags," *Washington Post* (March 30, 1988), p. F3.
93. "Air Bag Bonus," *Automotive News* (April 4, 1988), p. 6.
94. Quote from Joan Claybrook, "Airbags on the Way as Chrysler Gives In," *New York Times* (May 26, 1988), p. D4.
95. "Air Bags Slated for First Chrysler Models in May," *Automotive News* (February 8, 1988), p. 1.
96. *New York Times* (June 21, 1988), pp. A14–15.
97. "Air Bags on the Way as Chrysler Gives In," *New York Times* (June 21, 1988), p. 1.
98. Interview of Brian O'Neill, July 20, 1988.
99. "Air Bags on the Way as Chrysler Gives In," p. 1.
100. Ford Press Release, Washington, D.C., June 22, 1988.
101. Joseph B. White, "GM Will Offer Air Bag Devices on Some 1990 Cars," *Wall Street Journal* (July 14, 1988), p. 6.
102. Ibid.
103. *Public Citizen* vs. *Diane Steed*, U.S. Court of Appeals of the District of Columbia, July 15, 1988 (slip opinion), p. 13.
104. Ibid.
105. Brian O'Neill, "Ford Is First Again on Air Bags—Soon Passengers as Well as Drivers Protected," News Release, Insurance Institute for Highway Safety, Washington, D.C., June 22, 1988, p. 1.

Chapter 10

ASSESSING AMERICA'S PERFORMANCE

The U.S. Congress declared in 1966 that the "overriding goal" of the National Traffic and Motor Vehicle Safety Act was to reduce deaths and injuries resulting from automobile crashes. A new administrative agency was established and granted expansive authority to enact minimum safety standards for all new cars sold in the United States. The legislative history of the act indicates that Congress intended the federal government to emphasize standards that would mitigate injuries arising from the so-called "second collision" between motorists and the vehicle interior.

A dispassionate observer in 1966 had good reason to be skeptical about the probable effectiveness of the radical reform of auto safety policy. New regulatory legislation did not alter the basic calculations of the marketplace. Motorists were not demanding much safety ("it won't happen to me"), and leaders of the auto industry were convinced that safety was a poor marketing strategy ("safety doesn't sell"). Although the new National Highway Safety Bureau was intended to alter these traditional perceptions, the anticipated obstacles to effective regulatory policy were formidable.

NHSB (and later NHTSA) was a small bureau buried within a new Cabinet-level department. The overall mission of the Cabinet secretary was transportation, not safety. To regulate under the act, the bureau had to recruit a professional staff from scratch except for a few transfers from the small safety office in the Department of Commerce. Congress never authorized enough funding to allow NHTSA to hire more than several hundred professional employees, and from the beginning the agency had difficulty hiring first-rate engineers and scientists. In particular, NHTSA's budgetary

219

and technical resources were no match for its regulatee—the largest manufacturing industry in the world. Detroit had more money, more expertise, better access to data, and the potential—as yet unrealized—to wield awesome political power in Washington.

Nor was NHTSA the sole power center in Washington on auto safety. The NHTSA administrator reported to the secretary of transportation, who possessed authority to modify or block policy initiatives arising from NHTSA. Likewise, NHTSA policy could be reviewed by officials at the White House and the Office of Management and Budget. Even if the executive branch were united in support of ambitious auto safety regulation, opponents might still pull off a blocking strategy on Capitol Hill or in the federal courts.

The 1970s and 1980s were not a hospitable era for major new regulatory programs aimed at big business and American motorists. The spiraling "misery index" (defined as the sum of the rates of inflation, interest, and unemployment) reduced the political appeal of regulatory programs aimed at American business in general and Detroit in particular. Meanwhile, the burgeoning conservative movement in American politics made it difficult for Washington to impose tough new regulatory policies on states, businesses, and private citizens. During this same period it became apparent that public confidence in regulation had deteriorated. The so-called consumer "movement" of the 1960s proved to be more of a tentative "impulse" than a profound grass-roots revolution in American politics.[1]

The Record of Accomplishment

In light of the improbability of success, America has achieved significant progress in its thirty-year struggle to increase occupant-crash protection through regulation. By 1988 the progress was reflected in several of tangible outcomes: installation of manual restraint systems in all seating positions of new motor vehicles; improvements in the comfort and convenience of manual systems; legislation in fifty states requiring parents to restrain their infants and toddlers in cars; a transformation in public attitudes toward safety belts; legislation in thirty-two states requiring adult motorists to use manual safety belt systems; a substantial increase in the use of manual restraint systems by motorists of all ages; installation of innovative automatic belt systems and improved air bags in an increasing percentage of new cars; and significant reductions in the frequency and severity of injuries from crashes. The progress in each of these areas has been substantial, even though the pace has been frustratingly slow.

Front-seat lap belts were installed as standard equipment in all new cars beginning with the 1964 model year in response to state legislation. Rear-seat lap belts were added as standard equipment in model years 1965 and 1966, again as an anticipatory response to political demands by state legislators. A separate shoulder harness requirement was imposed by NHSB beginning January 1, 1968. This requirement was amended by NHTSA in the early 1970s to require a combination lap/shoulder belt system that could be fastened with a single buckling action. Combination lap/shoulder belts for *rear-seat* occupants are now being offered by several car manufacturers and will probably become standard equipment in all new cars by the early 1990s.[2]

The nation's economic investment in manual restraint systems is not trivial. Two front-seat and two rear-seat belt systems are estimated to cost $150 per car.[3] At a sales rate of ten million new cars per year, the total national investment in manual belt systems is about $1.5 billion dollars per year. Such an investment is potentially worthwhile, for numerous studies show that occupants restrained by lap/shoulder belts can reduce their risk of fatal and serious crash injuries by 40 to 50 percent.

By itself, the installation of manual restraint systems in cars did little for safety. Throughout the 1970s, only a small fraction of American motorists—fewer than one-fifth—were buckling up. Even though Australia, Canada, and many European countries required adults to buckle up in the 1970s, the United States seemed preoccupied by an elusive quest for passive restraint systems.

Whereas most developed nations took steps to restrain adults in cars before requiring young children to do so, the United States took the opposite course. Thanks to the efforts of some activist pediatricians and medical societies, America emerged as the world's pioneer in child-restraint legislation. Beginning with Tennessee in 1978 and ending with Wyoming in 1985, every state in the nation adopted legislation requiring the use of child-restraint devices. As the observational data in Table 10–1 indicate, legislation in America was accompanied by a steady increase in the proportion of infants and toddlers restrained during highway travel.

The transformation in public opinion about safety belts for adults did not occur until the mid-1980s, when adults began to accept the necessity of mandatory belt use legislation for themselves. Table 10–2 reveals the extent of this change in public attitudes, which is associated with the onset of Elizabeth Dole's creative regulatory compromise. Pollster Gary Lawrence characterizes this trend as "one of the most phenomenal shifts in attitudes ever measured by pollsters."[6]

Table 10–1   Percentage of Children (Infants and Toddlers) Observed Traveling Unrestrained in 19 American Cities, 1979–1986

Year	Percent Unrestrained
1979	83.1
1980	78.4
1981	73.5
1982	65.3
1984	47.1
1985	37.6
1986	22.3
1987	15.3

SOURCE: National Highway Traffic Safety Administration, *Restraint System Usage in the Population* (Washington, D.C.: U.S. Dept. of Transportation, various years).

Table 10–2   Public Attitudes Toward Mandatory Belt Use Legislation

Question: "Would you favor or oppose a law that would fine a person $25 if he or she did not wear a seat belt when riding in an automobile?"

Year	Favor	Oppose	No Opinion
1972	23	71	6
1977	17	78	5
1982	19	75	6
1985	35	59	6
1988	54	43	3

SOURCE: George Gallup, Jr., and Alec Gallup, "Seat Belt Use Quadruples Since 1982," *The Gallup Poll* (June 26, 1988), p. 1.

In 1984 New York became the first state in the nation to enact comprehensive belt use legislation for drivers and passengers of all ages. A domino effect ensued. By 1988, thirty-two states comprising about three-fourths of the U.S. population had enacted mandatory belt use laws of some form.[7] Although the citizens of Nebraska and Massachusetts repealed their laws at the polls in November 1986, these actions have only been replicated by one other state.[8]

Table 10–3   Self-Reported Belt Use in the United States

Question: "Thinking about the last time you got into a car, did you use a seat belt, or not?"

Year	Percent "Yes"
1973	28
1977	22
1982	17
1984	25
1985	40
1986	52
1987	65
1988	69

SOURCE: George Gallup, Jr., and Alec Gallup, "Seat Belt Use Quadruples Since 1982," *The Gallup Poll* (June 26, 1988), p. 1.

Concerted efforts are still in progress to persuade state legislators in the remaining sixteen states to enact belt use laws.[9]

Changes in public attitudes toward safety belts were accompanied by changes in behavior. From 1970 to 1980 the rate of safety belt use among motorists in the United States was persistently low, less than 20 percent based on numerous roadside surveys,[10] and actually declined during the 1970s from 20 to about 10 percent in 1980.[11] National rates of belt use were raised several percentage points by 1985 due to increased public awareness of safety and grass-roots educational campaigns aimed at promoting belt use.[12] When belt use legislation in states began to take effect in the 1985–1987 period, belt use rates climbed quickly and dramatically. Nationally, the rate of observed belt use among drivers increased from 15 percent in 1984 to 40 percent in 1987.[13] Similar trends, based on opinion surveys, are summarized in Table 10–3.

Among states with belt use laws, the average belt use rate hovers just below 50 percent.[14] In states that practice aggressive primary police enforcement (e.g., North Carolina and Texas), belt use rates are commonly observed in excess of 60 percent.[15] The highest rates of belt use are observed in the first few months after laws take effect; then belt use declines to a plateau that is typically two to three times larger than the rate of belt use observed in the prelaw period.[16] Traffic safety experts have linked the new laws and higher belt use rates to reductions in the incidence of nonfatal injuries and fatalities. The injury prevention estimates from various studies differ, but the pattern is clear.

In the state of New York, fatalities and serious injuries among front-seat occupants declined by 20 and 13 percent, respectively, in the first nine months after the legislation took effect in January 1985.[17] A combined study of New York, Michigan, New Jersey, and Illinois found that postlaw occupant fatality rates were 5 to 16 percent lower than prelaw fatality rates.[18] Another study of eight states with belt use laws found that laws with primary enforcement reduced occupant fatalities by about 10 percent whereas laws permitting only secondary enforcement reduced such fatalities by about 7 percent.[19] The favorable effects of belt use legislation on nonfatal injuries have also been demonstrated through analysis of insurance claims information in New York and New Jersey.[20] By the fall of 1987, one analyst estimated that belt use legislation in America had already prevented more than 1,000 fatalities and 100,000 nonfatal injuries.[21]

The 1980s also brought progress on the passive-restraint front. NHTSA's automatic restraint regulation stimulated car manufacturers to make available to new-car buyers a variety of innovative automatic belt designs. In model year 1987, three basic types of automatic belts were offered:

1. Motorized shoulder belts (detachable or nondetachable) with knee bolsters (and/or manual lap belts) such as the designs offered on the Toyota Cressida and the Ford Escort.
2. Nonmotorized shoulder belts and knee pads (and/or manual lap belts) such as the designs offered on the Volkswagen Jetta and the Chrysler Dodge Daytona.
3. Nonmotorized automatic lap and shoulder belts (detachable) such as the designs offered on GM's Buick Skylark and Pontiac Bonneville.

Roughly 1.4 million cars were equipped with automatic belts in model year 1987, or about 14 percent of 1987 sales. About half of the systems are motorized and half nonmotorized.[22]

The rate of belt use by motorists in voluntary usage environments varies dramatically as a function of belt design. Manual lap/shoulder belts are worn by 15 to 30 percent of motorists (absent legal requirements).[23] Automatic lap/shoulder belts (nonmotorized) are worn by roughly 50 percent of motorists.[24] Nonmotorized and motorized shoulder belts achieve usage rates of about 70 and more than 90 percent, respectively.[25]

Preliminary studies suggest that automatic belts are effective injury-prevention systems. One study found 27 percent fewer injury insurance claims on 1983–1984 Toyota Cressida models (with motorized shoulder belts) than on 1983–1984 Nissan Maximas (with manual lap/shoulder belts). Collision coverage losses for the two cars were comparable, suggesting that the lower rate of injury claims is indeed attributable to belt use.[26] Earlier studies of Volkswagen's automatic shoulder belt and knee pads were also encouraging. The overall frequency of insurance claims for occupant injuries in 1981–1982 VW Rabbits with automatic seat belts was 14 percent lower than for comparable Rabbits with manual lap/shoulder belt systems. The difference was not accounted for by corresponding differences in the frequency of collision losses.[27] More precise estimates of the safety benefits of automatic belts must await evaluation of the large-scale field experience now being accumulated by 1987 and later models.

Air bags were not offered to car buyers in America from 1970 to 1984 except for the 10,000 GM luxury cars equipped with full-front air bags during model years 1974–1976. More recently, Mercedes-Benz sold about 170,000 cars (1984–1986) equipped with driver-side air bags in the United States, and Ford sold about 11,000 cars with driver-side air bags in 1985 and 1986.

During the first year of NHTSA's automatic restraint rule (model year 1987), a total of approximately 154,000 cars were sold with air bags.[28] Driver-side systems were marketed by Mercedes-Benz, BMW, Volvo, Ford, and Honda. A full-front air bag system was also offered by Porsche. The best estimates are that air bag offerings will expand rapidly between model years 1988 and 1994. Ford, GM, and Chrysler are planning to offer almost five million cars with driver-side air bag systems by the early 1990s with Chrysler the first to offer them as standard equipment in May 1988. In 1988 Ford became the first domestic manufacturer to announce plans to offer full front-seat air bags as standard equipment.

The early field experience with the improved air bag systems of the 1980s has been good. The devices have deployed when they were supposed to and, unlike the 1974–1976 GM system, they have not deployed inadvertently in significant numbers. The precise injury-mitigation effects of air bags cannot yet be assessed because cars with air bags have not been driven enough on the road. The good news is that the large-scale, real-world experience with the air bag that has been needed since the early 1970s is now taking place. By 1990 the value of modern air bag designs should be much clearer to everyone.

Ingredients of Success

The major achievements in occupant crash protection from 1950 to 1988 have been widespread installation of manual belt systems, adoption of mandatory restraint use laws for adults and children, and the development and large-scale field testing of passive-restraint devices, such as automatic belt designs and air bags. Because this success story was quite improbable, it is instructive to look back and pinpoint the key ingredients of the success.

The basic epidemiological and biomechanics research of the 1940s and 1950s laid the scientific foundation for the political break-throughs. Crash-injury researchers demonstrated that the mechanical energy associated with high-speed car crashes need not cause severe injury or death to car occupants. The "second collision" was eminently survivable as long as the vehicle interior was designed in a forgiving fashion.

Creative science by itself would not have assured safety progress. The marketplace was not receptive to safety innovation without governmental regulation. The formidable barriers to effectively marketing safety are evident from Ford's safety campaign of 1956, the history of safety belts in the 1960s, GM's experience with optional air bags in the mid-1970s, and Ford's experience with optional driver-side air bags in the mid-1980s.[29] America has learned that if it wants dramatic improvements in car safety, the latent prosafety sentiments in the marketplace will have to be supplemented by a strong political demand for safety. Moreover, advocates of new safety innovations must prove that the innovation is safe, reliable, and effective enough to justify its monetary and nonmonetary costs to consumers.

In 1966 the U.S. Congress made just such a demand. Responsibility for radical reform legislation lies primarily with congressional entrepreneurs, LBJ, media fascination, and Ralph Nader. The swiftness and far-reaching character of the legislation arose in part out of the outrage associated with the ill-advised decision of certain GM officials to spy on Ralph Nader.

Radical legislation and a new federal regulatory agency were not sufficient to assure that optimally safer cars would be built and sold or that motorists would take advantage of safety features. Implementation of the auto safety agenda was pursued for two decades by a cadre of career officials and political appointees at the National Highway Traffic Safety Administration and the U.S. Department of Transportation. Every NHTSA administrator from William Haddon to Diane Steed contributed to development of the air bag technology. Numerous career NHTSA officials have devoted their professional lives to devising, justifying, and defending regulations that are partly responsible for both the manual lap/shoulder belt systems now in all cars and the innovative automatic restraint systems now being marketed by Detroit. Whenever NHTSA's commitment to safety regulation seemed to weaken, consumer activists and insurers exposed the change in policy and sought judicial or congressional correction.

The opposition of Detroit's safety engineers to unrealistic passive-restraint deadlines was also an ingredient of the success story. If the relatively primitive, untested air bag systems of the early 1970s had been installed on an industry-wide basis, adverse public reactions might have led to premature rejection of the technology. Industrial opposition bought precious time for early air bag systems to be tested, refined, and transformed into commercially acceptable products. The extra time also facilitated the development of innovative automatic belt designs that are proving to be quite popular among some new-car buyers. In this unrecognized fashion, critics of the air bag proved to be engineers of safety progress.

Throughout the 1970s and early 1980s the Insurance Institute for Highway Safety and its sponsors in the insurance industry played a critical role in keeping the air bag on the national political agenda. IIHS scientists filled NHTSA's rule-making docket with pro–air bag evidence and arguments, while IIHS media professionals planted pro–air bag stories in newspapers, popular periodicals, talk shows, and television news programs. Working in conjunction with consumer and public health groups, IIHS also kept key congressmen and senators informed about what was happening to the air bag.

An independent federal judiciary performed a remarkably constructive role. The *Chrysler* court (1972) affirmed NHTSA's regulatory authority while allowing time for GM scientists to complete their pathbreaking work on instrumented dummies for crash tests. Without the dummies, the safety of air bag deployment could not be tested and compliance with an automatic restraint standard could only by judged by a "rubber yardstick." The *Pacific Legal* court (1978) avoided taking sides in political objections to NHTSA's passive-restraint rule and thereby pushed politics into the appropriate forum: the U.S. Congress. The Supreme Court's *State Farm* decision (1983) was a model of judicial independence: upholding NHTSA's obligation to pursue safety when congressional and executive branch actors had become apathetic due to economic hard times. By keeping safety on the national political agenda, the Supreme Court provided a new opportunity for creative political compromise.

Elizabeth Hanford Dole performed the role of master pluralist. She saw that coercion of Detroit on the air bag issue was a losing strategy—politically and technologically. She saw that mandatory seat belt laws, the most promising short-run policy, had no determined advocates. She saw that the insurance industry was determined to make sure that automatic restraints received the large-scale, real-world test that had been neglected for fifteen years. She also wanted the Reagan administration to do something better than reinstate the Carter administration's policy. Faced with these constraints, she devised a plan that accomplished what had appeared to be impossible. She turned virtually all interested parties into advocates of both mandatory seat belt laws and automatic restraints. Through compromise Elizabeth Dole created a powerful consensus in favor of safety progress.

By itself the Dole plan did not save the air bag technology. A new generation of progressive executives at Ford Motor Company followed the European lead of Mercedes-Benz and turned to the air bag in their search for long-run competitive advantage. By successfully bargaining with their historical adversaries in the insurance industry, Ford manipulated the regulatory process in its favor and triggered a domino effect in the industry. GM, Chrysler, and the Japanese were left scrambling to defend themselves against Ford's progressive image on auto safety. At Ford, air bags and motorized belts are no longer viewed as methods of regulatory compliance per se. They are now elements of a long-run corporate strategy to build Ford's reputation as the manufacturer that supplies a quality product that customers learn to appreciate.[30]

Tactical Errors

Idealists might argue that the successes were too little, too late. How does one justify the years of bickering, delay, and lost opportunity? Perhaps a different political-economic system or regulatory framework would have accomplished more safety, sooner. In a sense the idealists are right. One can easily imagine scenarios that might have brought automatic restraints and belt use legislation to America more readily and swiftly. One must remember, however, that democratic pluralism, unlike authoritarianism, is intended to be a highly deliberate and inefficient approach to problem solving.

Even so, a good case can be made that America's failure to achieve more rapid safety progress is not a reflection of flaws in American pluralism. Tactical errors by regulators and corporate executives who sought to defy the logic of pluralism contributed to much of the conflict, wasted time, and lost opportunities. While any political-economic system must expect a degree of managerial incompetence, a case can be made that a keener understanding of the logic of pluralism *by the participants in this saga* would have helped accelerate safety progress.

In the early 1970s, John Volpe and Douglas Toms may have slowed the commercialization of air bags by playing the technology-forcing game too rigidly. Ford and Chrysler were driven to other power centers in search of relief from what they saw to be incompetent rule making, and they found that NHTSA's rigid position was vulnerable. The Nixon White House and the federal courts might have been less sympathetic to the complaints of Ford and Chrysler if NHTSA had adopted a careful phase-in plan such as the one proposed by President Ed Cole of General Motors Corporation.

Likewise, NHTSA Administrator James Gregory may have erred in 1975 by refusing to embrace George Eads's suggestion of a large-scale field test of air bags. By pushing forward with an ambitious regulatory proposal, Gregory set in motion a rule making that Transportation Secretary Coleman, the Ford administration, and the auto industry were not prepared to accept. Ultimately, Secretary Coleman negotiated a test program similar to the one proposed by Eads. By then, however, the Ford administration had lost control of the issue to the incoming Carterites. If the demonstration had been initiated two years earlier by Gregory, it would have been much more difficult for the Carter administration to overturn because major capital investments would have already been made.

In retrospect, the decision of Adams to overturn the Coleman plan also appears to have been a mistake. Obviously, the benefit of hindsight is enormous, especially given the unpredictability of the energy situation in 1977. Moreover, the commitments of manufacturers to produce air bags under the Coleman plan may not have been legally binding. Nevertheless, the commitments were certainly political ammunition in the event that Detroit reneged. Once such commitments were replaced by a performance standard, Claybrook was left with little recourse when the industry abandoned air bags in favor of detachable automatic belts. As a political personality, Claybrook would have been much more credible and effective as an enforcer of Detroit's demonstration program commitments than as a defender of the reasonableness of an industry-wide regulation during tough economic times.

The unraveling of Senator Warner's compromise plan in 1980 is a clear case of stakeholders putting ideology before safety progress. Senator Howard Metzenbaum and Ralph Nader knew that the Warner plan would do more for safety than would the incoming

Reagan administration. By refusing to compromise, they may have delayed for five years the resolution of the controversy. In their defense, late-night conference committee negotiations are not a sensible forum for making car safety regulations. It is also clear now that Secretary Dole's compromise was more powerful than Warner's plan because Dole also created strong incentives for the enactment of mandatory belt use laws.

Administrator Peck's approach to the automatic restraint rule was a clear tactical error. The deregulation strategy was poorly defended and not accompanied by a viable nonregulatory policy to make passive restraints available to consumers. More importantly, the decision by Peck reflected a lack of appreciation for the fact that insurers and consumer activists—let alone career NHTSA officials—had a strong stake in the decision and deserved to have a hand in any new policy that might be fashioned. By trying to defy the pluralist principle that there are multiple sources of power, Peck set up the Reagan administration for a major defeat in the Supreme Court. It is remarkable that a public servant as talented as Drew Lewis would have allowed Peck to make such an implausible decision. At a minimum, Peck's rationale needed to address explicitly the technology that was the major source of political controversy: the air bag.

Peck's error arose from an even more serious error in corporate policy by the leadership of the domestic auto industry. Roger Smith, Henry Ford II, and Lee Iacocca thought they could win the fight against mandatory passive restraints through exercise of presidential power in the form of "regulatory relief." By opposing regulation on the grounds that they would comply by installing easily detached automatic belts, Detroit belittled the regulatory process. When Detroit refused Peck's call for an air bag demonstration, they left Peck blowing in the wind. The Supreme Court saw Detroit's position in 1981 for what it was: an industry-wide attempt to obstruct the regulatory process in a fashion inconsistent with NHTSA's safety mandate from Congress.

Perhaps the most persistent error—one that reflects misguided strategy as much as poor tactics—was the failure of all of the major stakeholders to promote mandatory belt use legislation in the 1970s. Career NHTSA officials were preoccupied with engineering solutions and gave only sporadic support to belt use legislation. The leadership of the domestic auto industry was too lethargic to make the major political investment in lobbying state legislatures that was required. Insurers and consumer activists mistakenly perceived belt use laws as a threat to passive restraints. Scientists at the Insurance Institute for Highway Safety contributed to the myopia by evaluating belt use education solely on the narrow terms of whether or not it persuaded people to buckle up. They didn't acknowledge that some programs may have had a broader political value as a device to soften public opposition to mandatory belt use legislation.

The administrators of NHTSA during the 1970s are especially vulnerable to the criticism that they failed to aggressively promote manual belt use. They fell victim to the self-fulfilling prophecy that mandatory belt-use legislation was politically infeasible. As recently as 1984, Joan Claybrook wrote that "it is not likely that mandatory seat belt usage laws will be either enacted or found acceptable to the public in large numbers because of objections to the discomfort and inconvenience of existing belt systems."[31] What Elizabeth Dole demonstrated in the 1980s was that NHTSA could—through grass-roots organizing and cooperation with the private sector—play an influential role in determining whether such legislation was politically viable.

Despite these tactical and strategic errors, there were several excellent examples of successful administrative and corporate leadership throughout the saga. In terms of administrative leadership, DOT's John Volpe and NHTSA's Douglas Toms showed that a "technology-forcing" strategy, if pursued with realism, can stimulate the creative engineering talents of Detroit, Europe, and Japan. DOT's William Coleman showed how regulatory power can be used as a chip in bargaining with auto companies to promote safety innovations. NHTSA's Joan Claybrook used political skills to protect passive-restraint regulation from legislative veto, to seek progressive legislative compromise in the face of fierce antiregulation sentiments, and to help kill the "trapdoor" in Sacramento. Master pluralist Elizabeth Dole pulled it all together in a coherent administrative strategy that assured safety progress in one form or another.

There were also several examples of successful corporate strategy. Volkswagen's development of a viable alternative to the air bag in the early 1970s obtained for the firm a relatively secure position despite persistent regulatory uncertainties. Chrysler and Ford won the battle over air bags with GM's Ed Cole in 1972 by exposing NHTSA's political, technical, and legal vulnerabilities. State Farm Insurance Company had no difficulty in distinguishing stockholder interest from political ideology in its successful lawsuit against the Reagan administration in the early 1980s. Ford, though, displayed the most innovative strategy when it went to Washington and obtained a creative pro–air bag agreement with its arch foe the insurance industry.

Incremental Reforms

Although much of the delay in America's resolution of the occupant-restraint controversy can be attributed to strategic and tactical errors by participants, there were also basic weaknesses in the regulatory system that accentuated inefficiencies in decision making. Several of these weaknesses can be corrected or minimized through incremental reform of the administrative process. Other weaknesses must be accepted as the price America pays for the considerable virtues of democratic pluralism.

Perhaps the most profound weakness of the current process is the lack of continuity in public policy that our electoral system fosters. Vacillation reflects the give and take of partisan American politics. As presidents and congressional coalitions change, attitudes toward safety policy change. Some will argue that this lack of continuity in public policy is healthy because it is democracy at work. Others see vacillating safety policy as foregone opportunity.

Vacillation in America's policy toward car safety is aggravated by the short length of tenure that characterizes political appointees to the U.S. Department of Transportation and the National Highway Traffic Safety Administration. Table 10–4 reveals that the average DOT secretary serves in the job for less than three years, while the average NHTSA administrator serves for just greater than three years. These averages have been increased considerably by the relatively long tenures of Elizabeth Dole at DOT and Diane Steed at NHTSA.

The high turnover rate among leaders of DOT and NHTSA is especially disruptive to car safety regulation because of the long lead times that characterize product development in the auto industry. It now takes at least five years to plan, design, engineer, and

Table 10–4   Length of Tenure of DOT Secretaries and NHTSA Administrators (1967–1989)

Secretary of Transportation		NHTSA Administrator	
Alan Boyd	(1967–1969)	William Haddon	(1966–1969)
John Volpe	(1969–1972)	Douglas Toms	(1969–1972)
Claude Brinegar	(1973–1975)	James Gregory	(1973–1976)
William Coleman	(1975–1976)	John Snow	(1976–1977)
Brock Adams	(1977–1979)	Joan Claybrook	(1977–1980)
Neil Goldschmidt	(1979–1980)	Raymond Peck	(1981–1983)
Drew Lewis	(1981–1983)	Diane Steed	(1983–1989)
Elizabeth Dole	(1983–1987)		
James Burnley	(1987–1988)		

tool a new model. Add-on safety features can be included with less lead time, but fundamental technology-forcing rules require at least five years lead time.

The mismatch between the time horizons of political appointees in Washington and safety engineers in Detroit weakens the ability of NHTSA to accomplish fundamental advances in occupant-crash protection. Each political leader is inclined to look for programs with short-run safety payoffs while neglecting attention to long-run design improvements. A new leader at NHTSA has little incentive to expend limited agency resources putting the finishing touches on a predecessor's ambitious new car safety rule.

There is no panacea to this problem since the time horizon of American politicians is inherent to our electoral process. Modest steps could nonetheless be taken to strengthen the hand of career officials in policy making at NHTSA. As we have seen, career public servants tend to have strong loyalties to agency mission and informal ties with congressional sponsors. They are also likely to have gained some respect for the long-run mission of the agency. By strengthening the power of civil servants relative to political appointees, the interests in continuity of agency policy might be better served.

Several modest steps could be taken in this direction without crippling the ability of the NHTSA administrator to change policy when appropriate. The number two job at NHTSA, the deputy administrator, could be transformed into a career position that would rotate among senior agency officials at the discretion of the NHTSA administrator. A tradition could be established that the deputy administrator publishes a separate opinion on major rule-making initiatives, amendments, and rescissions. Such a voice for career people in the agency could be added to the rule-making record and thereby made available for congressional oversight and judicial review. A good case can be made that the major blunders in NHTSA's twenty-year history—the 1974 starter-interlock rule and the 1981 rescission of the automatic-restraint rule—would not have occurred if career agency officials had played a stronger role in the decision-making process. Such an arrangement would not prevent political appointees from differing from career officials but it would induce them to take a stronger interest in the decision preferences of senior agency officials. The hurdles to hasty changes in long-run agency policy would thereby be heightened.

Another fundamental weakness in modern administrative government is the secrecy that often characterizes communications between the White House and the leadership of DOT and NHTSA. Certainly the White House (or its agents in OMB) should take responsibility for channeling NHTSA's energies in a direction that is consistent with both safety and other national policy objectives such as economic growth and energy independence. A safety-oriented agency such as NHTSA can also benefit from rigorous analytical review of its policy initiatives by professional economists at OMB. No noble purpose is served, however, by censoring White House and OMB communications from the rule-making record and public view. For example, the intervention of John Ehrlichman in Volpe's rule making on behalf of President Nixon should have been "on the record." Likewise, the communications between Chris DeMuth of OMB and Raymond Peck and James Burnley in the Reagan years should have been "on the record." If the White House has ideas to contribute, they should be conveyed as an explicit contribution to the pluralist competition. Any ideas that are too shallow or ill-motivated to be publicized don't need to be a part of the process.

Because the American political style relies on trust to foster bargaining and compromise, the administrative process must be honest and open to everyone. Secrecy breeds mistrust and thereby fosters unresolvable conflict. The exercise of power through secrecy is also dangerous because it facilitates autocratic decisions that cannot be effectively checked by interest groups at other power centers. The starter-interlock requirement is an example of a regulatory policy that appeared to evolve from an antipluralist process: A secret meeting of Ford executives with the Nixon White House was followed by a secret meeting between Ehrlichman and Volpe that was followed by a sharp policy reversal at NHTSA. It is no accident that the starter-interlock rule proved to be one of the most counterproductive policies in the history of automobile safety regulation.

A model case of White House oversight occurred during the Ford administration when George Eads of the Council on Wage and Price Stability recommended publicly that NHTSA launch a large-scale demonstration of the air bag technology. NHTSA officials were irritated that COWPS did not endorse a regulatory strategy, but they could not argue with the process. The COWPS recommendation was made openly and was therefore susceptible to debate and modification. When NHTSA chose to reject this suggestion, it did so ironically to the detriment of the technology and safety mission the agency was supposedly promoting.

Finally, the process of regulating auto safety in America is permeated by suspicion and lack of trust, even on basic matters of science and engineering. Disputes about facts and values are hopelessly intermingled. There was no institutional player in the thirty-year saga that built a reputation for providing sound, objective information and advice on the key technical questions of concern to regulators.

Some degree of adversarialism is of course inevitable and appropriate because of the conflicting wants of interested parties. And it is not necessarily possible to devise an institution that is completely disinterested and unbiased. While trustworthy science and engineering are therefore an ideal, they are nonetheless an ideal worthy of pursuit.

A model of such an institution, in the environmental science area, is beginning to build a reputation for both scientific integrity and responsiveness to societal needs. The Health Effects Institute (HEI) in Cambridge, Massachusetts, is a nonprofit corporation whose mission is to provide sound scientific information on the potential health effects of emissions from motor vehicles. HEI is jointly and equally funded by the U.S. Environmental Protection Agency and the world's major vehicle and engine manufacturers. Though it is too early to tell how successful HEI will be, a promising start has already been made and the institutional model is an intriguing one.

An HEI-style institution could have exerted a constructive influence throughout the occupant-restraint controversy. There were numerous technical controversies about restraint systems that begged for dispassionate analysis:

- The effectiveness of air bags in fatality and injury mitigation.
- The reliability and validity of the dummies used in crash tests.
- The risk of injury to out-of-position children from air bag deployment on the passenger side.
- The design of proper emergency release mechanisms for passive belt systems that allow an occupant to exit from the vehicle after a crash.
- The estimated costs to consumers of various passive restraint systems.

As long as these kinds of technical issues are shrouded in uncertainty and distrust, the public debate will necessarily remain at a low level of sophistication. Once these issues are clarified in an objective fashion, the debate will then turn to the key value differences that are truly driving the controversy. In a democratic society, it is useful for citizens to grapple with their value differences so that avenues of compromise can be explored.

Conclusion

Much has been written about the supposed failure of modern health, safety, and environmental legislation. The regulatory legacy spawned by Ralph Nader and environmentalism has supposedly been captured by business interests, throttled by insufficient congressional appropriations, and/or rendered ineffective by technical and administrative complexities. In light of the above, why did the struggle for occupant crash protection through restraints prove to be at least partially successful? I conclude with three speculative hypotheses that require further examination in other case studies of social regulation.

First, the emergence of the insurance industry as a potent prosafety, proregulation interest group was a critical factor. The insurance industry does not necessarily match the auto industry in either political or economic clout, but its resources are clearly enormous. Hence, the American public does not have to rely solely on poorly financed activist groups on the political left to carry the burden of advocacy. Such activist groups are often the only proregulation forces in consumer and environmental disputes. Indeed, the mere existence of a prosafety economic titan may sometimes act as a deterrent to obstructionist corporate strategies.

Second, the benefits of auto safety regulation are more tangible, compelling, and susceptible to scientific documentation than is the case for many other forms of health, safety, and environmental regulation. All of us have known relatives or friends who were injured or killed in car crashes. Likewise, the statistical frequency of such events makes it possible for analysts to link effective regulations to changes in the frequency of such events. In contrast, the health benefits of environmental control—while perhaps real—are more uncertain and much more difficult to prove. When the health benefits of regulation are perceived to be uncertain or elusive, antiregulation forces can more effectively manipulate the multiple power centers in American politics to serve their interests.

Finally, NHTSA's single-industry focus and the oligopolistic structure of the auto industry discourage corporate strategists from engaging in persistent antiregulation behavior. Because NHTSA is unlikely to be abolished and manufacturers know they must deal with NHTSA on numerous issues, they are more likely to try to exploit NHTSA's powers for competitive advantage than to try to undermine NHTSA's powers. In contrast, most social regulatory agencies (say, EPA, OSHA, CPSC, and FDA) cover multiple industries that are more classically competitive than the auto industry. No single firm is likely to manipulate such a diverse agency for competitive purposes, and many such firms will go years without facing a serious regulatory question. In this setting, no single firm has an incentive to invest heavily in firm-specific regulatory strategies, while all firms have a modest incentive to reduce uncertainty by marshalling an antiregulation industry strategy. A testable implication of this hypothesis is that industry-wide associations should be more active in, say, EPA rule makings than in NHTSA rule makings. In this regard, it is interesting to note how modest a role the Motor Vehicle Manufacturers Association played throughout the twenty-year auto safety saga. In any event, regulators who face several large regulatees in one industry may face an easier task promulgating rules than regulators who face numerous regulatees in diverse industries.

Regardless of the precise explanations, the process of auto safety regulation in America progressed in accordance with the basic tenets of democratic pluralism. People with conflicting interests and values battled for thirty years in multiple power centers. The controversy was persistent, intense, and at times ugly. Although the searches for some kind of resolution often seemed fruitless, a light is in fact at the end of the tunnel.

Endnotes

1. Michael Pertschuk, *Revolt Against Regulation: The Rise and Pause of the Consumer Movement* (Berkeley: University of California Press), 1982.
2. "'Ford Phasing in Rear Restraints," *Automotive News* (July 20, 1987), p. 31.
3. Barry Felrice, personal interview, July 23, 1987.
4. John D. Graham and Max Henrion, "A Probabilistic Analysis of the Passive-Restraint Question," *Risk Analysis* 4:25–40 (1984).
5. Alexander C. Wagenaar, "Mandatory Child Restraint Laws: Impact on Childhood Injuries Due to Traffic Crashes," *Journal of Safety Research* 16:9–21 (1985).
6. Geoff Sundstrom, "Seat Belts Advance from a Trend to a Habit, Study Finds," *Automotive News* (February 15, 1988), p. 120.
7. Charles Spillman, personal interview, September 15, 1987.
8. "Mandatory Seat Belt Laws Voted Down in Two States," *Automotive News* (November 17, 1986), p. 64.
9. "Safety Group Seeks to Hike Belt Use to 70% by 1990," *Automotive News* (September 14, 1987), p. 40.
10. Ann Grimm, "Use of Restraint Systems: A Review of the Literature," *The HSRI Review* 11:11–28 (1980).
11. *Safety Belt Use Among Drivers*, Final Report of Opinion Research Corporation to U.S. Department of Transportation, NHTSA, Washington, D.C., DOT-HS-805-398, May 1980.

12. *Restraint Use in the Traffic Population,* U.S. Department of Transportation, NHTSA, Washington, D.C., DOT-HS-806-714, March 1985.

13. "39% Buckle Up, NHTSA Reports," *Automotive News* (9 February 1987), p. 98.

14. B. J. Campbell and Frances A. Campbell, *Seat Belt Law Experience in Foreign Countries Compared to the United States* (Falls Church, Va.: AAA Foundation for Traffic Safety, December 1986).

15. Ibid.

16. Ibid.

17. Eric A. Latimer and Lester B. Lave, "Initial Effects of the New York State Auto Safety Belt Laws," *American Journal of Public Health* 77:183–186 (1987).

18. Adrian K. Lund, Paul Zador, and Jessica Pollner, "Motor Vehicle Occupant Fatalities in Four States with Seat Belt Use Laws," SAE Technical Paper Series Number 870224.

19. Alexander C. Wagenaar, "Effects of Mandatory Seat Belt Laws on Traffic Fatalities in the United States," Transportation Research Institute, University of Michigan, Ann Arbor, Michigan, May 1987.

20. *A Preliminary Evaluation of New York and New Jersey Insurance Injury Claim Results Before and After Enactment of Mandatory Seat Belt Legislation, 1983–85 Models,* Highway Loss Data Institute, A-27, Washington, D.C., May 1986.

21. B.J. Campbell, personal interview, September 15, 1987.

22. *Industry and Consumer Response to New Federal Motor Vehicle Safety Requirements for Automatic Occupant Protection,* Report to Congress, U.S. DOT, NHTSA, May 1987.

23. *Restraint System Use in the Traffic Population,* U.S. Department of Transportation, NHTSA, DOT-HS-806-987, May 1986.

24. "Drivers Find It's a Cinch with Ford's New Automatic Belt System," *Status Report: Highway Loss Reduction* (June 27, 1987), pp. 1–4; "Ford Leads GM in Use of Passive Belts," *Automotive News* (June 15, 1987), p. 94.

25. *Automatic Safety Belt Systems: Owner Usage and Attitudes in GM Chevettes and VW Rabbits,* U.S. Department of Transportation, NHTSA, Washington, D.C., DOT-HS-805-399, May 1981; *Automatic Safety Belt Usage in 1981 Toyotas,* U.S. Department of Transportation, NHTSA, Washington, D.C., DOT-HS-806-146, February 1982.

26. "Study Gives Nod to Passive Belts," *Automotive News* (December 1, 1986), p. 19.

27. *Injury Claim Frequencies for Volkswagen Rabbits with Automatic and Manual Seat Belts,* Highway Loss Data Institute, Washington, D.C., Research Report HLDI A-21, April 1984.

28. *Industry and Consumer Response,* May 1987.

29. "Ford Air Bags Get Low Showroom Rating," *Automotive News* (September 14, 1987), p. 34.

30. " 'Safety Doesn't Sell?' That's Nonsense, Ford Says," *Automotive News* (August 3, 1987), p. 57.

31. Joan Claybrook, *Retreat from Safety* (New York: Pantheon Books, 1984), pp. 186–187.

Appendix

Interview Sources

Eugene Ambroso, safety engineer, General Motors Corporation

Milford Bennett, assistant director, auto safety engineering, General Motors Corporation

B. J. Campbell, Director, Highway Safety Research Center, University of North Carolina

Robert L. Carter, retired career NHTSA employee

Joan Claybrook, former administrator, NHTSA

Greg Dahlberg, Staff, House Appropriations Committee

Christopher DeMuth, former Director of Information and Regulatory Affairs, Office of Management and Budget

Lowell Dodge, Former Director, Center for Auto Safety

Cindy Douglas, former staff member, Commerce Committee, U.S. Senate

Howard Dugoff, former career NHTSA employee and deputy administrator of NHTSA

George Eads, former assistant director for government operations, COWPS

Barry Felrice, career NHTSA employee

Michael Finkelstein, career NHTSA employee

Merrick Garland, attorney, Arnold and Porter

James Gregory, former administrator, NHTSA

Roy Haeusler, retired former director of auto safety engineering, Chrysler Corporation

Philip W. Haseltine, former Deputy Assistant Secretary for Policy and International Affairs, DOT

Ralph Hitchcock, career NHTSA employee

James Hofferberth, career NHTSA employee

Thomas Hopkins, former staff economist, Office of Management and Budget

Charles Hurley, Vice President, Public Policy, National Safety Council

Benjamin Kelley, former vice president, Insurance Institute for Highway Safety

Christopher Kennedy, former director, federal government affairs, Chrysler Corporation

Joseph Kennebeck, Washington representative, Volkswagen of America, Inc.

Charles Livingston, former career NHTSA employee

David Martin, retired former director, auto safety engineering, General Motors Corporation

Roger Maugh, former automotive safety director, Ford Motor Company

Don McGuigan, senior attorney, Ford Motor Company

Carl Nash, career NHTSA employee

Brian O'Neill, president, Insurance Institute for Highway Safety

Raymond Peck, former administrator, NHTSA

Helen Petrauskas, vice president of environmental and safety engineering, Ford Motor Company

Robert Rogers, director, auto safety engineering, General Motors Corporation

George Smith, retired General Motors Corporation safety specialist

Jerome Sonosky, attorney, Hogan and Hartson

Charles Spilman, president, Traffic Safety Now

William Steponkus, Steponkus and Associates

Judith Stone, Director of Government Affairs, National Safety Council

Douglas Toms, former administrator, NHTSA

Malcolm Wheeler, attorney, Skadden, Arps, Slate, Meagher and Sloan

Published Background Sources

Automotive Industries
Automotive News
Best's Review
Business Week
Congressional Quarterly Almanac
Fortune
Industry Week
Iron Age
Machine Design
New York Times
Product Engineering
Status Report: Highway Loss Reduction
Wall Street Journal
Ward's Automotive Report

INDEX